T0202934

Lecture Notes in Computer Science 13428

More information about this series at https://link.springer.com/bookseries/558

Robert Wrembel · Johann Gamper ·
Gabriele Kotsis · A Min Tjoa ·
Ismail Khalil (Eds.)

Big Data Analytics and Knowledge Discovery

24th International Conference, DaWaK 2022
Vienna, Austria, August 22–24, 2022
Proceedings

Editors
Robert Wrembel 🄳
Poznań University of Technology
Poznań, Poland

Johann Gamper 🄳
Free University of Bozen-Bolzano
Bozen-Bolzano, Italy

Gabriele Kotsis
Johannes Kepler University of Linz
Linz, Austria

A Min Tjoa
Vienna University of Technology
Vienna, Austria

Ismail Khalil
Johannes Kepler University of Linz
Linz, Austria

ISSN 0302-9743　　　　　　ISSN 1611-3349 (electronic)
Lecture Notes in Computer Science
ISBN 978-3-031-12669-7　　ISBN 978-3-031-12670-3 (eBook)
https://doi.org/10.1007/978-3-031-12670-3

This Springer imprint is published by the registered company Springer Nature Switzerland AG
The registered company address is: Gewerbestrasse 11, 6330 Cham, Switzerland

Preface

DaWaK was established in 1999 as the International Conference on Data Warehousing and Knowledge Discovery. It ran continuously under this name until its 16th edition (2014, Munich, Germany). In 2015 (Valencia, Spain) it was renamed the International Conference on Big Data Analytics and Knowledge Discovery to better reflect new research directions in the broad and dynamically developing area of data analytics, but it retained its DaWaK acronym. In 2022, the 24th edition of DaWaK took place during August 22–24 in Vienna, Austria, back as a physical event following online editions in 2020 and 2021.

Since the very beginning, the DaWaK conference has been a high-quality forum for researchers, practitioners, and developers in the field of data integration, data warehousing, data analytics, and, recently, big data analytics. The main objectives of this event are to explore, disseminate, and exchange knowledge in these fields through scientific and industry talks. Big data analytics and knowledge discovery remain hot research areas for both academia and the software industry. They are continuously evolving, fueled by advances in hardware and software. Important research topics associated with these major areas include data lakes (schema-free repositories of heterogeneous data), conceptual/logical/physical database design, theoretical foundations for data engineering, data integration (especially linking structured and semi-structured data sources), big data management (mixing relational tables, text, and any types of files), query languages (beyond SQL), scalable analytical algorithms, parallel storage and computing systems (cloud, parallel database systems, Spark, MapReduce, HDFS), graph processing, stream and time series processing, IoT architectures, artificial intelligence/machine learning algorithms, and applications of these solutions in industry.

DaWaK 2022 attracted 57 papers, from which the Program Committee selected 12 regular papers and 12 short papers, yielding an acceptance rate of 21% for the regular paper category, of 27% for the short paper category, and of 42% overall. Each paper was reviewed by at least three reviewers and in some cases up to four. The accepted papers cover a variety of research topics on both theoretical and practical aspects. The program included the following topics: (1) text analytics, (2) data warehousing and OLAP, (3) feature selection algorithms, (4) time series processing, (5) schema discovery and construction, (6) pattern discovery, and (7) machine learning algorithms. Thanks to the reputation of DaWaK, selected best papers of DaWaK 2022 will be invited for a special issue of the Data & Knowledge Engineering (DKE, Elsevier) journal. Therefore, the PC chairs would like to thank the DKE Editor-in-Chief, Carson Woo, for his approval of the special issue.

We would like to express our sincere gratitude to all Program Committee members and the external reviewers who reviewed the papers thoroughly and in a timely manner. Finally, we would like to thank the DEXA conference organizers for their continuous support and guidance especially Ismail Khalil for providing a great deal of assistance (as always), putting his experience at our disposal.

August 2022   Johann Gamper
                                                                      Robert Wrembel

Organization

Program Committee Chairs

Robert Wrembel	Poznan University of Technology, Poland
Johann Gamper	Free University of Bozen-Bolzano, Italy

Steering Committee

Gabriele Kotsis	Johannes Kepler University Linz, Austria
A Min Tjoa	TU Wien, Austria
Robert Wille	Software Competence Center Hagenberg, Austria
Bernhard Moser	Software Competence Center Hagenberg, Austria
Alfred Taudes	Vienna University of Economics and Business and Austrian Blockchain Center, Austria
Ismail Khalil	Johannes Kepler University Linz, Austria

Program Committee

Alberto Abello	Universitat Politècnica de Catalunya, Spain
Cristina D. Aguiar	USP São Carlos, Brazil
Syed Muhammad Fawad Ali	Poznan University of Technology, Poland
Witold Andrzejewski	Poznan University of Technology, Poland
Sylvio Barbon	University of Trieste, Italy
Ladjel Bellatreche	LIAS/ENSMA, France
Sadok Ben Yahia	University of Tunis El Manar, Tunisia
Fadila Bentayeb	ERIC Lab, University of Lyon, France
Jorge Bernardino	ISEC - Polytechnic Institute of Coimbra, Portugal
Vasudha Bhatnagar	University of Delhi, India
Sandro Bimonte	INRAE, France
Pawel Boinski	Poznan University of Technology, Poland
Kamel Boukhalfa	University of Science and Technology Houari Boumediene, Algeria
Omar Boussaid	ERIC Lab, University of Lyon, France
Stephane Bressan	National University of Singapore, Singapore
Wellington Cabrera	University of Houston, USA
Sharma Chakravarthy	University of Texas at Arlington, USA
Silvia Chiusano	Politecnico di Torino, Italy
Alfredo Cuzzocrea	University of Calabria, Italy

Laurent D'Orazio	University of Rennes, CNRS, IRISA, France
Jérôme Darmont	Université Lyon 2, France
Soumyava Das	Teradata Labs, USA
Karen Davis	Miami University, USA
Claudia Diamantini	Università Politecnica delle Marche, Italy
Alin Dobra	University of Florida, USA
Christos Doulkeridis	University of Piraeus, Greece
Markus Endres	University of Passau, Germany
Leonidas Fegaras	University of Texas at Arlington, USA
Philippe Fournier-Viger	Shenzhen University, China
Matteo Francia	University of Bologna, Italy
Filippo Furfaro	University of Calabria, Italy
Pedro Furtado	University of Coimbra, Portugal
Luca Gagliardelli	University of Modena and Reggio Emilia, Italy
Enrico Gallinucci	University of Bologna, Italy
Kazuo Goda	University of Tokyo, Japan
Maria Teresa Gómez López	University of Seville, Spain
Marcin Gorawski	Silesian University of Technology and Wroclaw University of Technology, Poland
Sven Groppe	University of Lübeck, Germany
Frank Höppner	Ostfalia University of Applied Sciences, Germany
Mirjana Ivanovic	University of Novi Sad, Serbia
Stephane Jean	LISI/ENSMA and University of Poitiers, France
Selma Khouri	Ecole Nationale Supérieure d'Informatique, Algeria
Uday Kiran	University of Tokyo, Japan
Jens Lechtenbörger	University of Münster, Germany
Young-Koo Lee	Kyung Hee University, South Korea
Carson Leung	University of Manitoba, Canada
Sebastian Link	University of Auckland, New Zealand
Sofian Maabout	LaBRI, University of Bordeaux, France
Patrick Marcel	Université de Tours, LIFAT, France
Alejandro Maté	University of Alicante, Spain
Paolo Missier	Newcastle University, UK
Jun Miyazaki	Tokyo Institute of Technology, Japan
Anirban Mondal	University of Tokyo, Japan
Rim Moussa	ENICarthage, Tunisia
Kjetil Nørvåg	Norwegian University of Science and Technology, Norway
Boris Novikov	National Research University Higher School of Economics, Russia
Makoto Onizuka	Osaka University, Japan

Carlos Ordonez	University of Houston, USA
Jaroslav Pokorný	Charles University in Prague, Czech Republic
Praveen Rao	University of Missouri, USA
Franck Ravat	IRIT, Université de Toulouse, France
Oscar Romero	Universitat Politècnica de Catalunya, Spain
Keun Ho Ryu	Chungbuk National University, South Korea
Ilya Safro	University of Delaware, USA
Abhishek Santra	University of Texas at Arlington, USA
Kai-Uwe Sattler	Ilmenau University of Technology, Germany
Alkis Simitsis	HP Labs, USA
Günther Specht	University of Innsbruck, Austria
Kostas Stefanidis	Tampere University, Finland
Emanuele Storti	Università Politecnica delle Marche, Italy
Olivier Teste	IRIT, France
Dimitri Theodoratos	New Jersey Institute of Technology, USA
Maik Thiele	HTW Dresden, Germany
Juan Trujillo	University of Alicante, Spain
Maurice van Keulen	University of Twente, The Netherlands
Panos Vassiliadis	University of Ioannina, Greece
Isabelle Wattiau	ESSEC and CNAM, France
Lena Wiese	Goethe University Frankfurt, Germany
Szymon Wilk	Poznan University of Technology, Poland
Yinuo Zhang	Teradata Labs, USA
Yongjun Zhu	Yonsei University, South Korea

External Reviewers

Nabil El Malki
Chiara Forresi
Joseph Giovanelli
Alex Mircoli
Uchechukwu Njoku
Shivika Prasanna
Arun George Zachariah

Organizers

Contents

Time Series Processing

Schema Discovery and Construction

Pattern Discovery

Text Analytics

An Integration of TextGCN and Autoencoder into Aspect-Based Sentiment Analysis

Yi-Hang Tsai, Chia-Ming Chang, Kun-Hsiang Chen, and San-Yih Hwang[✉]

Department of Information Management, National Sun Yat-sen University, Kaohsiung, Taiwan
d094020001@nsysu.edu.tw, syhwang@mis.nsysu.edu.tw

Abstract. Due to the rapid increase in User-Generated Content (UGC) data, opinion mining, also called sentiment analysis, has attracted much attention in both academia and industry. Aspect-Based Sentiment Analysis (ABSA), a subfield of sentiment analysis, aims to extract the aspect and the corresponding sentiment simultaneously. Previous works in ABSA may generate undesired aspects, require a large amount of training data, or produce unsatisfactory results. This paper proposes a Graph Neural Network based method to automatically generate aspect-specific sentiment words using a small number of aspect seed words and general sentiment words. It subsequently leverages the aspect-specific sentiment words to improve the Joint Aspect-Sentiment Autoencoder (JASA) model. We conduct experiments on two datasets to verify the proposed model. It shows that our approach has better performance in the ABSA task when compared with previous works.

Keywords: Aspect-based sentiment analysis · Attention mechanism · Autoencoder · Text graph convolution network · Text mining

1 Introduction

With the rapid growth of the Internet and mobile devices, people can easily share their opinions on the Web. There are plenty of sources that users may express their opinions and preferences on various venues, such as social media platforms, online forums, and E-commerce websites. As it is time-consuming to manually analyze sentiments from massive data, automatic approaches that make good use of these data become imperative. Sentiment analysis or opinion mining intends to solve the problem by measuring subjective opinions automatically. Intuitively, sentiment analysis can be defined as a classification problem, namely classifying documents (or sentences) into different sentiment polarities. Further works in this line of research consider not only sentiment but also the aspect pertaining to the sentiment in the documents (or sentences). This line of research is called aspect-based sentiment analysis (ABSA). For instance, consider the sentence, "The new iPhone 13 produces detailed images." We may conclude that the sentence

R. Wrembel et al. (Eds.): DaWaK 2022, LNCS 13428, pp. 3–16, 2022.
https://doi.org/10.1007/978-3-031-12670-3_1

addresses the aspect "camera" and is positive because of the existence of the aspect word "images" and the corresponding sentiment word "detailed." Figure 1 shows some further examples and their expected aspect/sentiment output from an ABSA model.

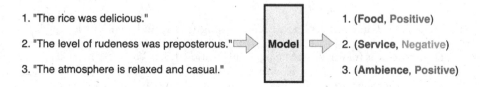

1. "The rice was delicious."

2. "The level of rudeness was preposterous."

3. "The atmosphere is relaxed and casual."

Model

1. (**Food, Positive**)

2. (**Service**, Negative)

3. (**Ambience, Positive**)

Fig. 1. An illustration of aspect-based sentiment model

Early approaches in ABSA used part of speech (POS) tagging and universal dependencies techniques to identify the associations between aspect words and sentiment words. For example, Hu & Liu [6] first compute the frequency of nouns or noun phrases in documents to extract aspect words and then employ universal dependencies to articulate sentiments by considering interconnected sentiment words and aspect words. Likewise, consider the sentence "The laptop price is very low." The adjective word "low" modifies the noun "price," which indicates a positive sentiment polarity of the price aspect. However, this method heavily depends on the accuracy of POS tagging and universal dependencies. It is observed that these two techniques are less accurate under certain situations, such as resource-light languages and casual writing.

Some approaches based on statistical models have been subsequently introduced to solve the ABSA problem. Brody & Elhadad [2] present a Local Latent Dirichlet Allocation (LDA) model to extract aspects, which works on sentences instead of documents. The Local LDA separates sentences into different aspects. A graph for each aspect is then constructed, where nodes are adjectives and connected by the co-occurrence in conjunctions. Specifically, a conjunction relationship is considered when two adjectives modify the same noun. Finally, sentiment propagation on the graph is used to determine sentiment polarities of words. Lin & He [8] proposed Joint Sentiment/Topic (JST) model, which is an extension of the LDA model and can detect document-level sentiment and extract a mixture of topics simultaneously. Jo & Oh [7] propose the Aspect and Sentiment Unification Model (ASUM), which is similar to JST. The main difference is that while JST allows the words of a document to be sampled from different word distributions, ASUM constrains the model such that the words from the same sentence must be sampled from the same word distribution. However, these approaches are unsuitable for sentences or short reviews due to the strong statistical constraint, resulting in a degraded performance in both aspect extraction and sentiment classification.

In recent years, neural networks have shown their capability in various machine learning applications, including ABSA. He et al. [4] use the autoencoder mechanism to find fine-grained aspects without supervision. Their model is shown to surpass LDA-based methods, yet desired aspects may not be generated by such an unsupervised method. Angelidis et al. [1] use a few aspect seed words as user supervision to extract aspects of a text segment (i.e., a sentence or a clause) and then employ a multiple-instance learning network model to learn a segment-level sentiment predictor using a hierarchical, attention-based neural architecture. Nevertheless, the process requires substantial sentiment ratings as training data. Zhuang et al. [12] build a joint aspect sentiment autoencoder model to predict aspect and aspect-specific sentiment. To supervise the model, they use aspect seed words and general sentiment seed words (e.g., good, bad, excellent, and terrible). Experimental results show that more accurate aspect-sentiment pairs can be obtained. However, we conjecture that better results may be achieved if the general sentiment seed words are replaced by aspect-specific ones. To do so, we propose a three-step approach in this work. The first step (Aspect Extraction) identifies aspect associated to each text segment. The second step (Aspect-based Sentiment Seed Word Generation) groups text segments of each aspect and applies Text GCN [11] to derive sentiment seed words for each aspect. In the final step, we modify the model proposed in [12] to take as input the aspect seed words and aspect-specific sentiment seed words. The experimental results confirm our conjecture and show that more accurate aspect-specific sentiments can be obtained. In summary, our work makes the following contributions:

- Our proposed approach, in contrast to many previous works, does not need supervision and labeled data.
- We propose using a graph convolution network to develop the aspect-specific sentiment words.
- Results on benchmark datasets (SemEval-2016) show that our approach improve the performance of aspect-based sentiment classification.

2 Methodology

We first describe the process of our approach and then provide details for each step. The architecture is shown in Fig. 2. We divide our approach into three steps: 1) aspect extraction, 2) aspect-based sentiment word generation, and 3) aspect sentiment detection. The aspect extraction step is to classify each sentence into aspects using the MATE model [1]. Next, we modify the Text GCN [11] to generate aspect-specific sentiment words. Finally, in the aspect-sentiment detection step, we propose an Aspect Specific Sentiment Autoencoder model (ASSA), which takes as input the aspect seed words and the aspect-based sentiment seed words obtained from the previous step. The ASSA model predicts both aspect and the associated sentiment for each sentence segment.

2.1 Problem Description

We denote C as a corpus of all reviews, in which each review consists of several sentence segments $(s_1, ..., s_m)$, and each sentence segment s_i contains a sequence

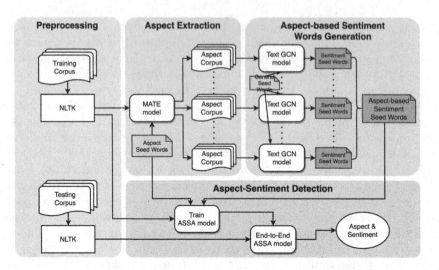

Fig. 2. Overall architecture of our approach

of words $(w_1^i, ..., w_n^i)$. Note that here we assume each sentence segment may involve zero or more aspects while conveying the same sentiment. For example, consider the sentence: "The room and the parking facility is nice, but the service is poor," which will be divided into two sentence segments: the first segment discusses room and parking aspects with positive sentiment, and the second segment focuses on the service aspect with negative sentiment. Let $V \in \mathbb{R}^{v \times d}$, where v is vocabulary size and d is vector dimension of each word, indicates the set of word vectors. We use $A_s \in \{0, 1\}^K$ to denote the aspects mentioned in a sentence s, where K is the number of aspects. Each sentence segment s have a polarity $pol_s \in [0, 1]$, where 0 and 1 indicate *Positive* and *Negative* sentiments respectively.

Our approach needs only a few aspect seed words and general sentiment words. For a given sentence segment s, we can predict the combination of its aspect labels and aspect-specific sentiment label (A_s, pol_s).

2.2   Aspect Extraction

To build a model for aspect extraction in an unsupervised way, we adopt the Multi-seed Aspect Extractor (MATE) model [1]. MATE is a weakly-supervised autoencoder for aspect extraction. In order to train the MATE model, some aspect seed words need to be given. We briefly illustrate the architecture of the MATE model which is shown in Fig. 3.

The MATE model is an autoencoder, which contains encoding and decoding steps. The Sentence Embedding z_s is transformed into the probability distribution over K aspects, specified by p_s in the encoding step. Pre-trained word embeddings at the Word Embedding layer are used to encode sentence embedding by attention mechanism. The MATE model makes the reconstructed embed-

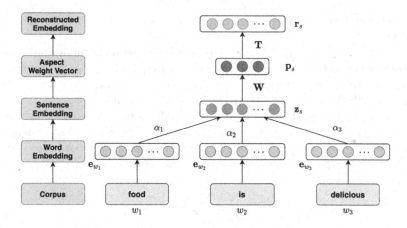

Fig. 3. Multi-Seed Aspect Extractor (MATE)

ding r_s similar to sentence embedding z_s in the decoding step by the linear combination of aspect embedding matrix $T \in \mathbb{R}^{K \times d}$ and the aspect distribution p_s. Note that each aspect embedding in T is obtained by weighted average of embeddings of aspect seed words.

2.3 Aspect-Based Sentiment Words Generation

With the MATE model obtained from the last step, we can classify all the sentence segments into aspects. Specifically, let p_s be the aspect distribution of a given sentence segment s. We conclude that s involves an aspect i if $p_s[i]$ is the maximum of all elements. From all the sentence segments involved in an aspect, we apply a graph convolution network approach, similar to the Text GCN [11], to obtain the sentiment seed words for that aspect. In detail, we build a graph for each aspect based on word co-occurrence and sentence-word relations, then learn the representations for both words and sentence segments by value propagation in the graph, as supervised by a few common sentiment words (e.g., good, bad, excellent, and terrible).

Formally, consider a graph $G = (V, E)$, where $V (|V| = n)$ and E are sets of nodes (i.e., words and sentences segments) and edges, respectively. In addition, X is a matrix representing features of nodes, namely the embedding of each word or sentence. The feature of a word is obtained from pre-trained word embedding, and the feature of a sentence segment is by averaging all word embeddings in a sentence segment. The adjacency matrix A of graph G describes the weights of edges. The goal is to derive the polarity Z for each word given X and A.

$$Z = f(X, A) = softmax(\tilde{A}ReLU(\tilde{A}XW^{(0)})W^{(1)}) \tag{1}$$

where $\tilde{A} = \hat{D}^{-\frac{1}{2}}\hat{A}\hat{D}^{-\frac{1}{2}}$ with $\hat{A} = A + I$ and I being the identity matrix, $\hat{D}$ is the diagonal node degree matrix of $\hat{A}$, and $W^{(l)}$ is a weight matrix for the l-th neural network layer. We set the matrix decomposition as the feature matrix of our nodes.

In the adjacency matrix A, an edge between a sentence node and a word node uses the term frequency-inverse document frequency (TF-IDF) value as its weight. The weight between word nodes is calculated using point-wise mutual information (PMI), a common measure for word associations [3].

In the original work [11], the Text GCN model is used to predict document-level sentiment, but in our work, the goal is to classify each word sentiment. We modify cross-entropy error over given general sentiment seed words (e.g., good, great, bad, and terrible). The loss function is defined as follows:

$$\mathcal{L} = - \sum_{s \in seed} \sum_{f=1}^{F} Y_{sf} ln Z_{sf}$$

Here, $F = 2$ represents the two polarities (positive and negative), and Z_{sf} is the predicted weight to sentiment f for the seed word s. Y_{sf} is the actual weight of sentiment of the seed word s.

After the population of the Text GCN model, we will obtain the polarity matrix Z, where each row contains the polarity vector of each word. To extract sentiment words of each aspect, we retain only verbs, adverbs, nouns, and adjectives based on Part-Of-Speech tagging. Then, the top 10% of frequent words are removed. Finally, we choose the top five words with the highest polarity per sentiment.

2.4   Aspect Sentiment Detection

Our goal is to identify the aspects and aspect-specific sentiment for a sentence segment. As the final step, we modify the Joint Aspect Sentiment Autoencoder (JASA) model [12] by employing the aspect-specific sentiment seed words instead of the general ones, and call the resultant model Aspect Specific Sentiment Autoencoder (ASSA). As illustrated in Fig. 4, the architecture of our ASSA model is similar to the JASA model, which can identify aspects and the corresponding sentiments. In the following, we will describe how we modify the aspect-specific sentiment reconstruction process.

The aspect and sentiment weight vectors are denoted p_s^A and p_s^S respectively. In order to generate aspect-specific sentiment, we need to derive the p_s^{AS} by taking the outer product of these two vectors and flattening it into a joint vector, as follows:

$$p_s^{AS} = vec(p_s^A \otimes p_s^S),$$

where $p_s^{AS} \in \mathbb{R}^{2K}$ describes the polarity of each of the two sentiments for each of the K aspects.

To reconstruct r_s^S from p_s^{AS}, we need to learn the embedding for each aspect-sentiment pair from the corresponding aspect-sentiment words obtained from the last step, collectively denoted $T^{AS} \in \mathbb{R}^{2K \times d}$. Finally, we can reconstruct r_s^S as follows:

$$r_s^S = T^{AS^{\mathrm{T}}} \cdot p_s^{AS}$$

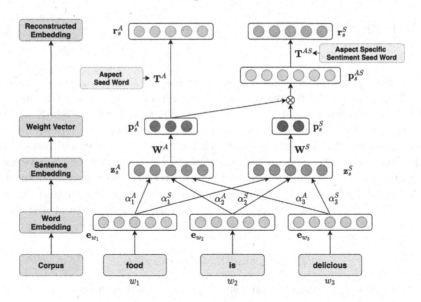

Fig. 4. Aspect Specific Sentiment Autoencoder (ASSA)

The JASA model places two regularizations, redundancy regularization, and seed regularization. However, in their result, redundancy regularization is not effective. So, we only use seed regularization to prevent each aspect (aspect t, $1 \leq t \leq K$, is represented by the t-th row of T^A) from deviating the weighted average of embeddings for given seed words. Accordingly, this seed regularization is the following:

$$C_A(\theta) = \sum_{t=1}^{K}[1 - sim(R^{A^{(t)}}, T^{A^{(t)}})],$$

where R^A denote the average embedding of seed words for each aspect. We use the same way to calculate this regularization term for the aspect-sentiment part, denoted as $C_{AS}(\theta)$.

In the objective function, besides reconstruction using loss and negative samples on the aspect, we also need to calculate the loss of sentiment prediction function and add to the regularization term. The final objective function is to minimize the total loss function that considers all above:

$$L(\theta) = L_A(\theta) + L_S(\theta) + \lambda(C_A(\theta) + C_{AS}(\theta))$$

$$L_A(\theta) = \sum_{s \in D} \sum_{i=1}^{m} \max(0, 1 - r_S^A z_S^A + r_S^A n_i)$$

$$L_S(\theta) = \sum_{s \in D} \sum_{i=1}^{m} \max(0, 1 - r_S^S z_S^S + r_S^S n_i),$$

where λ is the hyperparameter given to control the effect regularization term, and n_i is a negative sample.

3   Experiments

3.1   Dataset

Restaurant: The public Yelp dataset contains 8,635,403 reviews on various shops. We sample 50,000 reviews in the restaurant category for training our models in the restaurant domain. To evaluate our model, we use SemEval-2016 [10] in the restaurant domain as the test data set, which contains 1,120 sentences tagged to six aspects: ambience, drinks, foods, location, service, and restaurant. In addition, each sentence is tagged to one of the three sentiments: positive, neutral, and negative. To simplify the problem, we remove the aspect "restaurant" and the sentiment "neutral". As a result, we obtain a test dataset that has 937 sentences in the restaurant review domain. Statistics of the dataset are shown in Table 1.

Table 1. Statistics of test dataset

Domain	Aspect/Sentiment	Type	Number	Percentage
Restaurant	Aspect	Ambience	103	11.02
		Drink	45	4.81
		Food	462	49.41
		Location	15	1.6
		Service	312	33.39
	Sentiment	Positive	601	64.28
		Negative	336	35.94
Laptop	Aspect	Support	138	22.08
		OS	50	8.00
		Display	100	16.00
		Battery	77	12.32
		Company	80	12.80
		Mouse	54	8.64
		Software	67	10.72
		Keyboard	59	9.44
	Sentiment	Positive	276	44.16
		Negative	349	55.84

Laptop: Amazon product data [5,9] contain 7,824,482 reviews posted in the category of Laptops. We sample 50,000 reviews for training our models in the laptop domain. We also use SemEval-2016 in the laptop domain as the test data to evaluate our model. We select below eight entity types from all 21 types: Battery, Company, Display, Keyboard, Mouse, OS, Software, Support, and exclude "neutral" sentiment. Finally, we obtain a test dataset that has 625 sentences in the laptop review domain. Statistics of the dataset are given in Table 1.

3.2   Experimental Settings

We use NLTK to tokenize the restaurant and laptop domain training data and train word spaces respectively. The embedding size is set at 300, and the window size is 5. As for the parameters of the MATE model, the negative sample is set at 20 by default setting, and the learning rate is set at 10^{-5} for training smoothly. The aspect seed words used for the restaurant and laptop domains are shown in Table 2.

Table 2. Aspect seed word for both restaurant and laptop domains

Dataset	Aspect	Seed words
Restaurant	Ambience	Atmosphere, room, seating, environment, ambience
	Drink	Beverage, wines, cocktail, sake, drinks
	Food	Spicy, sushi, pizza, taste, food
	Location	Street, block, river, avenue, location
	Service	Tips, manager, waitress, servers, service
Laptop	Support	Service, warranty, coverage, replace, support
	OS	Windows, ios, mac, system, os
	Display	Screen, led, monitor, resolution, display
	Battery	Life, charge, last, power, battery
	Company	Hp, toshiba, dell, lenovo, company
	Mouse	Touch, track, button, pad, mouse
	Software	Programs, apps, itunes, photoshop, software
	Keyboard	Key, space, type, keys, keyboard

In the Text GCN model, the learning rate is set at 0.02, the number of units in the first hidden layer is set at 150, and the dropout rate is set at 0.5. All parameters follow the settings of the previous work [11]. The general sentiment words are identical in both domains: "good", "great", "nice", "best" and "amazing" for positive, while "gross", "bad", "terrible", "hate" and "disappointed" for negative. As for the parameter of our ASSA model, the negative sample is set at 20, the learning rate is set at 10^{-5}. The same setting used in MATE to ensure fair comparisons. According to previous work [12], the hyperparameter λ between 5 and 10 results in better performance, so here we set λ at 10.

3.3   Evaluation Metrics

For our classification task, we use four measures, including accuracy, precision, recall, and F1-score. To clarify, the aspect extraction is a multi-class task, and thus we apply macro-averaging on precision, recall, and F_1-score.

3.4 Experimental Results on Aspect Extraction

We compare our method with MATE [1] and JASA [12] on sentence-level aspect identification using both datasets. For each sentence s, each model returns an aspect vector (p_s or p_s^A). Since this vector is activated by softmax, we label the maximum class to represent the predicted aspect (A_s). The result is shown in Table 3. As can be seen, both JASA and ASSA perform better than MATE, which could be attributed to the training of aspect and sentiment together. In addition, ASSA has better performance in most cases, which shows the benefit of using aspect-specific sentiment seed words.

Table 3. Experiment results of aspect extraction (%)

Dataset	Method	Accuracy	Precision	Recall	F_1
Restaurant	MATE	68.98	65.97	55.66	55.60
	JASA	69.31	**67.58**	55.92	56.23
	ASSA	**69.73**	66.95	**56.19**	**56.32**
Laptop	MATE	75.36	74.32	75.00	74.09
	JASA	76.16	76.20	75.62	75.31
	ASSA	**76.96**	**77.34**	**76.48**	**76.50**

3.5 Experimental Results on Aspect-Sentiment Identification

For aspect-sentiment identification, we compare our method with the JASA model [12]. Additionally, we use a lexicon-based method as the baseline method by using MATE to extract aspects and text GCN to generate the sentiment lexicon. To be more precise, we use the aspect-specific sentiment words output by text GCN as the sentiment lexicon and account for the effect of negation words (e.g., not, never, less, and so forth). The sentiment of a sentence is obtained by aggregating the sentiment polarity of each constituent word. We name this method Word Sentiment Polarity (WSP) for brevity. We evaluate the performance of sentence-level sentiment identification on both Restaurant and Laptop datasets, and the result is shown in Table 3. For each sentence s, ASSA model returns an aspect-sentiment vector (p_s^{AS}). We select the aspect label from the aspect vector (p_s^A), then select the sentiment with the larger polarity score as the predicted sentiment (pol_s). We use four measures for the binary sentiment classification task (positive and negative): accuracy, precision, recall, and F1-score. As shown in Table. 4, both ASSA and JASA perform better than the baseline method WSP. Comparing ASSA with JASA, the aspect-specific sentiment words generated by Text GCN output by ASSA contributes to the improvement of sentiment classification accuracy on restaurant and laptop datasets by +1.8% and +2.1% respectively.

Table 4. Experiment results of sentiment identification (%)

Dataset	Method	Accuracy	Precision	Recall	F_1
Restaurant	WSP	74.18	72.66	74.18	72.97
	JASA	79.36	80.48	78.28	78.60
	ASSA	**81.18**	**81.84**	**79.84**	**80.34**
Laptop	WSP	65.28	61.86	68.74	59.95
	JASA	78.88	77.83	79.20	78.17
	ASSA	**80.96**	**79.84**	**81.59**	**80.26**

3.6 Discussion of Aspect-Based Sentiment Words

Table 5 and 6 show the sentiment seed words of each aspect generated by Text GCN. For each aspect in the restaurant domain, we list distinctive sentiment words. Notice that, we have "delicious", "love", "recommend", "definitely" noted as positive, and "disgusting", "worst", "awful", "inedible", "tasteless" noted as negative in the "Food" aspect, which are quite accurate. However, the Ambience and Drinks aspects have less significant words. We speculate that the performance of seed word generation by Text GCN might be related to the size of training data. As can be seen in Table 7 in Appendix, the number of sentences in the "Food" aspect accounts for about 40% of the entire training data, which is two to four times more than the others. Additionally, the two enormous (considering the size of training data) aspects in the Laptop domain also perform well, namely the Support and Display aspects. Even though other aspects seem less accurate, there are still some compelling cases. For example, the "macbook" and "mac" are considered positive in the OS aspect. This matches the stereotype that macOS has many advantages like stable, easy-to-use, and secure. Another example is in the Mouse aspect, where "freezes" and "jumps" are regarded as negative words, which are common problems related to computer mouses.

Table 5. The restaurant seed words generated by Text GCN

Aspect	Sentiment	Generated words
Ambience	Positive	Service, food, friendly, place, pad
	Negative	Dance, weird, completely, understand, head
Drinks	Positive	Food, service, awesome, recommend, ever
	Negative	Dry, rather, bottom, clearly, charged
Food	Positive	Delicious, also, love, recommend, definitely
	Negative	Disgusting, worst, awful, inedible, tasteless
Location	Positive	Food, love, favorite, place, always
	Negative	Filthy, dirty, broken, bouncer, smell
Service	Positive	Friendly, attentive, delicious, definitely, excellent
	Negative	Nasty, disgusting, rude, dirty, health

Table 6. The laptop seed words generated by Text GCN

Aspect	Sentiment	Generated words
Support	Positive	Fast, light, keyboard, little, price
	Negative	Worst, dealing, refused, sucks, defect
OS	Positive	Fast, macbook, mac, price, laptop
	Negative	File, driver, page, click, frustrating
Display	Positive	Price, fast, love, perfect, pros
	Negative	Bios, driver, install, drivers, update
Battery	Positive	Price, account, num, life, laptop
	Negative	Surface, lid, left, glossy, area
Company	Positive	Price, fast, recommend, product, thanks
	Negative	Repair, worse, worst, issue, send
Mouse	Positive	Pros, gaming, price, love, num
	Negative	Freezes, random, entire, jumps, cursor
Software	Positive	Fast, laptop, price, works, loves
	Negative	Norton, annoying, click, return, constantly
Keyboard	Positive	Love, easy, computer, light, laptop
	Negative	Horrible, support, return, bar, cursor

4 Conclusion

In this paper, we propose using the graph convolution network to improve the performance of the existing model [12] for the ABSA task. Our approach can be divided into three steps. The first step is to use the MATE model to extract aspects. For the second step, we propose to use Text GCN model to find aspect-specific sentiment words. The final step is training the ASSA model with aspect-specific sentiment seed words generated by the previous step to predict aspect and sentiment on given sentences. In the experiments, we show that it is effective in aspect-specific sentiment classification through our three-step approach. Moreover, our approach does not need supervision. Even though our proposed approach performs well for the ABSA task compared to other methods, the sentiment lexicon generated by Text GCN has room for improvement. Our future work includes improving Text GCN for generating better sentiment lexicon. In addition, it may be instructive to integrate the three steps into a single model for both efficiency and effectiveness.

Appendix: Statistics of Results from MATE Model

Table 7. Statistics of results from MATE model

Domain	Aspect	Number	Percentage (%)
Restaurant	Ambience	5166	10.33
	Drinks	5150	10.30
	Food	19942	39.89
	Location	9417	18.84
	Service	10319	20.64
Laptop	Support	8034	16.09
	OS	6366	12.75
	Display	8673	17.37
	Battery	5444	10.90
	Company	7125	14.27
	Mouse	4064	8.14
	Software	6845	13.71
	Keyboard	3380	6.77

References

1. Angelidis, S., Lapata, M.: Summarizing opinions: aspect extraction meets sentiment prediction and they are both weakly supervised. In: Proceedings of the 2018 Conference on Empirical Methods in Natural Language Processing, Brussels, Belgium, pp. 3675–3686. Association for Computational Linguistics, October-November 2018
2. Brody, S., Elhadad, N.: An unsupervised aspect-sentiment model for online reviews. In: Human Language Technologies: The 2010 Annual Conference of the North American Chapter of the Association for Computational Linguistics, pp. 804–812 (2010)
3. Church, K., Hanks, P.: Word association norms, mutual information, and lexicography. Comput. Linguist. **16**(1), 22–29 (1990)
4. He, R., Lee, W.S., Ng, H.T., Dahlmeier, D.: An unsupervised neural attention model for aspect extraction. In: Proceedings of the 55th Annual Meeting of the Association for Computational Linguistics (Volume 1: Long Papers), pp. 388–397 (2017)
5. He, R., McAuley, J.: Ups and downs: modeling the visual evolution of fashion trends with one-class collaborative filtering. In: Proceedings of the 25th International Conference on World Wide Web, pp. 507–517 (2016)
6. Hu, M., Liu, B.: Mining and summarizing customer reviews. In: Proceedings of the 10th ACM SIGKDD International Conference on Knowledge Discovery and Data Mining, pp. 168–177 (2004)

7. Jo, Y., Oh, A.H.: Aspect and sentiment unification model for online review analysis. In: Proceedings of the 4th ACM International Conference on Web Search and Data Mining, pp. 815–824 (2011)

8. Lin, C., He, Y.: Joint sentiment/topic model for sentiment analysis. In: Proceedings of the 18th ACM Conference on Information and Knowledge Management, pp. 375–384 (2009)

9. McAuley, J., Targett, C., Shi, Q., Van Den Hengel, A.: Image-based recommendations on styles and substitutes. In: Proceedings of the 38th International ACM SIGIR Conference on Research and Development in Information Retrieval, pp. 43–52 (2015)

10. Pontiki, M., et al.: SemEval-2016 task 5: aspect based sentiment analysis. In: Proceedings of the 10th International Workshop on Semantic Evaluation (SemEval-2016), San Diego, California, pp. 19–30. Association for Computational Linguistics, June 2016

11. Yao, L., Mao, C., Luo, Y.: Graph convolutional networks for text classification. In: Proceedings of the AAAI Conference on Artificial Intelligence, vol. 33, no. 01, pp. 7370–7377 (2019)

12. Zhuang, H., Guo, F., Zhang, C., Liu, L., Han, J.: Joint aspect-sentiment analysis with minimal user guidance. In: Proceedings of the 43rd International ACM SIGIR Conference on Research and Development in Information Retrieval, pp. 1241–1250 (2020)

OpBerg: Discovering Causal Sentences Using Optimal Alignments

Justin Wood[1]([✉]), Nicholas Matiasz[1], Alcino Silva[1], William Hsu[1], Alexej Abyzov[2], and Wei Wang[1]

[1] University of California, Los Angeles, CA 90095, USA
{juwood03,matiasz,silvaa,whsu,weiwang}@ucla.edu
[2] Mayo Clinic, 200 First Street, Rochester, MN 55905, USA
Abyzov.Alexej@mayo.edu

Abstract. The biological literature is rich with sentences that describe causal relations. Methods that automatically extract such sentences can help biologists to synthesize the literature and even discover latent relations that had not been articulated explicitly. Current methods for extracting causal sentences are based on either machine learning or a predefined database of causal terms. Machine learning approaches require a large set of labeled training data and can be susceptible to noise. Methods based on predefined databases are limited by the quality of their curation and are unable to capture new concepts or mistakes in the input. We address these challenges by adapting and improving a method designed for a seemingly unrelated problem: finding alignments between genomic sequences. This paper presents a novel method for extracting causal relations from text by aligning the part-of-speech representations of an input set with that of known causal sentences. Our experiments show that when applied to the task of finding causal sentences in biological literature, our method improves on the accuracy of other methods in a computationally efficient manner.

Keywords: Causality extraction · Sequence alignments · Zero-shot learning

1 Introduction

Researchers who perform biological experiments convey their discovery in published research articles, which contain descriptions of causal relations. This growing literature provides an enormous amount of information and represents the current state of biological understanding. This documentation of scientific discovery can verify previous experiments, provide insights to researchers [29], and motivate future research [23].

These corpora of biological text are growing at an exponential rate. Algorithms and approaches are thus needed to extract the relevant information, allowing biologists to understand and connect biological processes. Since researchers describe causal connections among biological entities in free-text research papers, it is logical to extract these connections using natural language processing (NLP).

A causal assertion can be thought of as a relation between an agent and a target. Often in biological studies, an agent is either passively observed or actively manipulated, and a change or lack thereof is noted in a target. Although this type of result can

R. Wrembel et al. (Eds.): DaWaK 2022, LNCS 13428, pp. 17–30, 2022.
https://doi.org/10.1007/978-3-031-12670-3_2

be described across many different and sometimes nonadjacent sentences, this paper focuses only on causal assertions appearing in a single sentence. This approach has the advantage of limiting the search range for descriptions of causality and takes advantage of existing methods that can reliably fragment documents into collections of sentences [22].

Existing methods for causality extraction use either predefined knowledge bases, word lists, other types of databases [6,17,25,27,33], or are based on statistical techniques—often some form of machine learning [3,8,11,20]. Predefined knowledge bases are of course limited by the quality of the knowledge base itself. Often, these sources are manually curated and do not always contain all possible words or phrases of interest. Additionally, they require exact matches to be useful. For instance, if a knowledge base contains causal verbs and a potential causal sentence contains the misspelled verb "cuases" (instead of "causes"), the sentence will be dismissed due to the misspelling. These predefined knowledge bases are also not able to capture new words or concepts, and they are not extensible to other tasks such as extracting causality from text in other languages.

One solution to these problems is to use existing machine learning techniques. But these approaches often require large amounts of labeled training data, something that can be expensive and tedious to obtain. These barriers of time and cost are expanded when the task is to discover more fine-grained details pertaining to causality, such as that of finding the specific types of studies and outcomes that lend evidence for a causal assertion. Additionally, the vocabulary for biomedical free text can be quite large, as it contains not only common words but also domain-specific terms. This large vocabulary set requires an even larger training data for the machine-learning model to predict the necessary components for representing causal phenomena.

Thus, to automatically extract causal sentences, an approach is required that does not suffer from limitations in the size of the training data, and that can be performed efficiently. The approach presented in this paper is inspired by the analogy of the aforementioned problem to that of comparing a set of genomic sequences in bioinformatics.

Though it may not be obvious, there is indeed a connection between aligning sequences in genomic data and finding causal sentences in free text. While each sentence may contain a unique set of words, the part-of-speech (POS) sequence of each sentence is likely to be much more common. Breaking each sentence into its grammatical structures can thus help to identify patterns in the way that causal relations are described. Thus, applying an alignment method to the grammatical structures of sentences has the potential to discover similarities that may be missed by approaches that focus only on words. We further illustrate this with the following example of three sentences and their corresponding POS mappings (for brevity we replace the POS label with a single character: P = pronoun, V = verb, D = determiner, A = adjective, N = noun, PP = preposition):

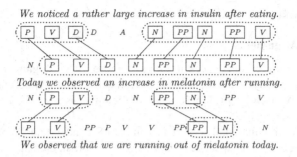

Here the first two sentences are talking about two different things; yet both are causal sentences. Their POS structures are similar. In comparison, the second and third sentence share a lot of words, more so than the first and second sentences, yet their POS representations have fewer matching elements, with long gaps in between matches. Therefore, knowing that the second sentence is causal, we cannot determine whether the third sentence is causal. It is our hypothesis that given a labeled set of causal sentences $C+$ and non-causal sentences $C-$, a new sentence s is classified as a causal sentence if its POS structure is most similar to a causal sentence and the similarity (S) is above a threshold δ,

$$\max_{c \in C+} S(c,s) > \max_{c \in C-} S(c,s) \wedge \max_{c \in C+} S(c,s) > \delta \qquad (1)$$

Our desired approach is to find causal relations by comparing the POS mappings of unlabeled sentences to that of labeled sentences. A new causal sentence is discovered by identifying the optimal number of alignments between the grammatical representations of the sentences. We show this alignment approach can thus classify causal sentences accurately and efficiently, and it has the potential to be used for other problems as well. The intuition behind this approach is rather simple: that people tend to describe causality using a similar grammatical form. In our alignment setting this similar grammatical form represents common subsequences—and the longer the common subsequence the more likely two compared sequences are making the same causal assertion.

Existing methods of sequence alignment are insufficient for aligning POS representations of free text: either (1) they require the user to specify the number of local alignments [1] or (2) they introduce a gap penalty for each new local alignment [16], possibly leading to erroneous alignments [1]. Given the nature of free text, it is unreasonable to ask the users to pre-specify the number of local alignments. Here, we generalize existing alignment algorithms by removing the need to specify these parameters, while keeping the same algorithmic complexity in terms of both space and time. This generalization allows us to efficiently apply the algorithm to text mining. The techniques presented in this paper need not be limited to extracting causality. We recommend using our approach for information retrieval tasks dealing with sequential similarity when the input data set is too small to be sufficient for machine learning.

2 Preliminaries

2.1 Sequence Alignments

Sequence alignment algorithms seek to assign a score for an alignment between two strings (A and B). The two most popular algorithms are Smith-Waterman [30] (local) and Needleman-Wunsch [24] (global). A local alignment is a maximal scoring alignment over the subsequences $A_p, A_{p+1}, A_{p+2}, \ldots, A_q$ and $B_x, B_{x+1}, B_{x+2}, \ldots, B_y$. A global alignment is the maximal scoring alignment over the A and B. Aligned strings often contain one or more instances of an *insertion* (or, interchangeably, *deletion*), which represents a single-character gap in the alignments. For example, with the alignments of the strings "BAT" and "BEAM" a global alignment could easily be:

global alignment:

```
B  –  A  T
|     |  |
B  E  A  M
```

		B	E	A	M
	0	-1	-2	-3	-4
B	-1	1	-1	-3	-4
A	-2	-1	0	0	-2
T	-3	-3	-2	-1	-1

with the "E" representing a deletion in the string "BEAM" and an insertion in "BAT". The "T" in "BAT" is aligned to the "M" in "BEAM"; since they are not the same, this is referred to as a *mismatch*. To find the alignments, most algorithms use dynamic programming with one or more two variable recurrence relations stored in a matrix. We demonstrate the matrix using the alignment of "BAT" and "BEAM" and scoring of match/mismatch $S(A_i, B_j) = (\pm) 1$ and indel score (Q) of -2 next to its the global alignment above.

2.2 AGE

AGE [1] is an alignment algorithm that addresses the problem of optimally aligning genomic sequences that contain large amounts of insertions or deletions; it addresses the problem that tuning the parameters of the Smith–Waterman [30] and Needleman–Wunsch [24] algorithms does not guarantee the optimal alignment of certain inputs. AGE solves this problem by introducing a maximum matrix into the recurrence that holds the maximum value of the equivalent location in the local alignment matrices. This approach, coupled with two local alignments—one going forward from the left end (L), and one going backward from the right end (R)—guarantees to locate the optimal "split point" of the two left and right local alignments. This algorithm guarantees finding the maximal left and right local alignments in quadratic time and space. Even though the space is polynomial, AGE can be unusable with a large input size. To address this, a linear space version can be formulated that achieves a memory bound of $\mathcal{O}(n)$ with computation time remaining at $\mathcal{O}(n^2)$.

2.3 Causality in Biological Literature

Much of biology is modeled as pathways, or signal cascades. Statements in the literature that describe causal relations between phenomena thus help biologists to understand these causal chains. And gaps in the understanding of a causal chain—i.e., its missing links—motivate hypotheses that then direct future research.

If an article is known to address the relation between two entities—e.g., a biological *agent*, A, and a biological *target*, T, then sentences in the article that mention both A and T are likely to describe that relation—for instance, in a results section that describes a study's empirical results, or in a discussion section that describes the study's implications for the field.

Beyond simply stating that two entities are causally related, a sentence can also state (1) the type of experiment that was performed on the two entities, and (2) the result of the experiment. One way to express this type of biological evidence is with the *research map* representation [29]: Each experiment is either an *intervention* (involving an experimental manipulation) or an *observation* (involving no manipulation); within these two classes, the change of the agent can be either positive or negative, yielding four experiment classes: *positive intervention* ($\uparrow$), *negative intervention* ($\downarrow$), *positive observation* ($\varnothing^{\uparrow}$), and *negative observation* ($\varnothing^{\downarrow}$). The results of these experiments are categorized as either *increase* ($+$), *decrease* ($-$), or *no-change* (0) to indicate how the target responded to a change in the agent.

2.4   Text Search and Extraction

A similar objective to causality discover is that of scientific search. This can be done using syntactic patterns [28], patters over dependency graphs [31] or using preexisting machine learning based methods [2, 10, 18]. Similarly, although not between scientific texts, search between parallel texts can be done effectively though word alignments [32]. This analysis also showed the applicability to corpora across different languages.

Also akin to the goals of causality extraction is that of relation extraction. The desiderata in relation extraction to discover relationships between different fragments in text. One technique to causal discovery without labeled data is using a predefined database [15]. Additionally, relation extraction has been shown to be inferred from side information [9]. Using this method results were obtained in zero-shot learning environment. Other methods utilize graph-based techniques [5] or neural networks to achieve superior improvements over baseline models [35].

Zero-shot learning can also be a setting in event extraction [21]. This allows extraction to be done in a way where annotation is not necessary—which is how event extraction is commonly done. In this setting transfer learning can be applied via neural networks to obtain results comparable to supervised methods.

3   Proposed Approach

The proposed approach, named *OpBerg*, builds upon the idea of the AGE algorithm: it uses a similar strategy to find the optimal number of local alignments. This optimal solution also uses our proposed concept of score length, whose definition is as follows:

Definition: Score Length. We define the score length for the alignment of POS tokens $a_i a_{i+1} \ldots a_{i+d_1}$ and $b_j b_{j+1} \ldots b_{j+d_2}$ as the difference between the max score in the alignment matrix at cell locations $(i + d_1, j + d_2)$ and (i, j). As an example the score length between "BA" and "AM" in Sect. 2.1 is $(-2) - (-3) = 1$.

A naive algorithm for solving the optimal alignment problem is to run the existing AGE method on every possible number of local alignments that could reasonably occur. To implement this approach, a new variable is introduced, k, which represents the current number of local alignments to run on the given input sequences. The results of these additions require an n factor increase in both running time and memory retention, where n is defined as the size of the largest input POS token sequence. The running time becomes $\mathcal{O}(n^3)$ with memory required as $\mathcal{O}(n^3)$.

It is intuitive to add a penalty (P) for each additional increase in local alignments. This penalty is needed since otherwise, the optimal alignment would always just match individual POS tokens. Because this penalty is proportional to the number of local alignments, we make the penalty a simple linear constant. The maximum alignment score can then be defined as:

$$\underset{1 \le k \le n}{Max} [P \times k + M(|A|, |B|, k)] \qquad (2)$$

where A and B are the input POS token sequences mapped from two sentences. M is the three-dimensional maximum matrix which holds the maximum alignment score for each a_i, b_j, and k; where $a_i \in A$ and $b_j \in B$.

A simple linear penalty constant reveals that returning one such alignment is not a trivial and deterministic task. The linear penalty can be thought of as an additional larger gap penalty, thus taking the form of a generalized global alignment [16]. It has already been shown [1] that this can lead to improper alignments.

The question then becomes: What is the optimal number of alignments? For example, a user may prefer to find an alignment that has only 1 large segment aligned and a score of 28 over 10 alignments and a score of 29. To determine the correct number of alignments, this work focuses on three major trade-offs:

1. Number of alignments.
2. Score length to break apart an alignment (α).
3. Min score length to start an alignment (β).

The naive algorithm solves the problem of finding the optimal number of local alignments, but it does so at a considerable cost. Opberg, the approach we present here, seeks to reduce memory by a factor of n and execution time by a factor of n.

3.1 OpBerg

Note that during execution of the naive algorithm described above, once it is decided that a new local alignment is a better choice, the optimal solution can then only be of the same or more alignments. This allows us to reuse the existing M matrix and shave off the k dimension, allowing for much simpler bookkeeping. We introduce a new matrix L that represents the values of a local alignment. The M matrix then takes on the interpretation of a matrix whose values are the max of the previous max M cell value and the corresponding L cell value. The optimal solution then can be in the L matrix (that is, performing a local alignment) or in the M matrix (that is, moving through the cells of the matrix and not decreasing in value). We use the notation that if the optimal solution is in the L matrix, then it is in the "L" or "alignment" state; and if the optimal solution is in the M matrix, then it is in the "M" or "max" state. Given that there is only one L state, it is entirely possible for the optimal solution to transition multiple times from the M state to the L state. We store the values of a transition in a new matrix N which holds the point of a transition in and out of the M state. Another matrix X holds the points of all transitions through the optimal solution.

The three trade-offs discussed above can be dealt with in various ways. To account for the number of alignments, we can leave in the original penalty P, but instead of considering this as a larger gap penalty, one can think of it as a value less than 1 and possibly even 0 (with the original gap penalty greater than 1). By doing so, one can easily gauge at what point a new alignment gap starts to weigh negatively on the score and thus becomes less desirable.

To consider the minimum score length needed to break apart an alignment, we need only to regard the point at which the algorithm exits the max state. If the current alignment has not dropped below the input score length α, then we will restrict the transition until the appropriate threshold has been reached.

We can apply a similar intuition as breaking apart an alignment to that of staring an alignment—with the change being the consideration of the entry into the max state (as opposed to the exit from the max state). We will restrict the length as we do for breaking apart an alignment, but a key difference happens when an alternative alignment is nonexistent. For example, a user may prefer not to start a segment of only 3 matched characters unless this is the max score out of any alternative alignments by a score of 3 matches. Therefore, we keep track of how a score length smaller than β influences the score. That is, we do not necessarily want to discard these alignments unless there is a better alignment available. A new parameter is introduced, $\gamma(x)$, which allows the user to specify a function to weigh how important a certain score length is when it is below threshold but no higher scoring alternatives exist.

With these parameters, the algorithm is bound to a running time of $\mathcal{O}(n^2)$ and memory requirements of $\mathcal{O}(n^2)$. The trade-off between the alignment score and number of jumps through the matrix to start a new local alignment is enforced by the penalty constants P, α, β, and function $\gamma(x)$.

Affine Gap. It should not always be the case that insertions and deletions (indels) between the inputs are weighted equally, regardless of where they occur. For instance, in certain causal sentences: a large cluster of indels may represent a tangential segment of words. To capture these occurrences, an affine gap model that takes into account segments of tangential words must be adapted to OpBerg.

For an affine gap, three matrices—representing a match/mismatch (L_G), insertion (L_I), and deletion (L_D) transitions, respectively—must be used in place of the original L matrix. The max matrix M cannot enter into any of these three states because it represents a jump through the inputs, so it remains the same. Also, since a local alignment must start and end with a match (diagonal move), the transition between the L states to the M states can occur only through the new L_G matrix. This also applies to the X and N matrices, as they only must monitor jumps between the L_G and the M matrices.

The recurrent relations needed for the affine gap OpBerg model are given in their entirety as (we take the shorthand p to represent the previous entry, i.e. $i_p = i - 1$):

$$L_I(i,j) = Max \begin{Bmatrix} L_I(i_p,j) + E \\ L_G(i_p,j) + F \\ L_D(i_p,j) + F \end{Bmatrix} \tag{3}$$

$$H_I(i,j) = \begin{cases} H_I(i_p,j) & \text{if } L_I(i,j) = L_I(i_p,j) + E \\ H_G(i_p,j) & \text{if } L_I(i,j) = L_G(i_p,j) + F \\ H_D(i_p,j) & \text{if } L_I(i,j) = L_D(i_p,j) + F \end{cases} \tag{4}$$

$$\delta(i,j) = Max \begin{Bmatrix} 0 \\ L_I(i_p,j_p) + S(a_i,b_j) \\ L_G(i_p,j_p) + S(a_i,b_j) \\ L_D(i_p,j_p) + S(a_i,b_j) \end{Bmatrix} \tag{5}$$

$$\psi(i,j) = \begin{cases} \theta(i,j) & \text{if } \delta(i,j) = 0 \\ H_I(i_p,j_p) & \text{if } \delta(i,j) = L_I(i_p,j_p) + S(a_i,b_j) \\ H_G(i_p,j_p) & \text{if } \delta(i,j) = L_G(i_p,j_p) + S(a_i,b_j) \\ H_D(i_p,j_p) & \text{if } \delta(i,j) = L_D(i_p,j_p) + S(a_i,b_j) \end{cases} \tag{6}$$

$$\epsilon(i,j) = \begin{cases} M(i_p,j_p) + S(a_i,b_j) + P & \text{if } \delta(i,j) - \psi(i,j) \le \alpha \\ -\infty & \text{otherwise} \end{cases} \tag{7}$$

$$L_G(i,j) = Max\{\delta(i,j), \epsilon(i,j)\} \tag{8}$$

$$H_G(i,j) = \begin{cases} Max\{M(i_p,j), M(i,j_p)\} & \text{if } L_G(i,j) = 0 \\ H_I(i_p,j_p) & \text{if } L_G(i,j) = L_I(i_p,j_p) + S(a_i,b_j) \\ H_G(i_p,j_p) & \text{if } L_G(i,j) = L_G(i_p,j_p) + S(a_i,b_j) \\ H_D(i_p,j_p) & \text{if } L_G(i,j) = L_D(i_p,j_p) + S(a_i,b_j) \\ Max\{M(i_p,j), M(i,j_p)\} & \text{if } L_G(i,j) = M(i_p,j_p) + S(a_i,b_j) + P \end{cases} \tag{9}$$

$$L_D(i,j) = Max \begin{Bmatrix} L_I(i,j_p) + F \\ L_G(i,j_p) + F \\ L_D(i,j_p) + E \end{Bmatrix} \tag{10}$$

$$H_D(i,j) = \begin{cases} H_I(i,j_p) & \text{if } L_D(i,j) = L_I(i_p,j_p) + F \\ H_G(i,j_p) & \text{if } L_D(i,j) = L_G(i,j_p) + F \\ H_D(i,j_p) & \text{if } L_D(i,j) = L_D(i,j_p) + E \end{cases} \tag{11}$$

$$\zeta(i,j) = \begin{cases} L_G(i,j) & \text{if } L_G(i,j) \ge \beta \\ \gamma(L_G(i,j)) & \text{otherwise} \end{cases} \tag{12}$$

$$M(i,j) = Max \left\{ \begin{array}{l} \zeta(i,j) \\ M(i_p,j) \\ M(i,j_p) \end{array} \right\} \tag{13}$$

$$X(i,j) = \begin{cases} X(i_p,j) & \text{if } L_G(i,j) = L_I(i_p,j) + S(a_i,b_j) \\ X(i_p,j_p) & \text{if } L_G(i,j) = L_G(i_p,j_p) + S(a_i,b_j) \\ X(i,j_p) & \text{if } L_G(i,j_p) = L_D(i,j_p) + S(a_i,b_j) \\ N(i_p,j_p) \cup (i,j) & \text{if } L_G(i,j) = \epsilon(i,j) \\ \varnothing & \text{if } L_G(i,j) = 0 \end{cases} \tag{14}$$

$$N(i,j) = \begin{cases} X(i,j) \cup (i,j) & \text{if } M(i,j) = \zeta(i,j) \\ N(i_p,j) & \text{if } M(i,j) = M(i_p,j) \\ N(i,j_p) & \text{if } M(i,j) = M(i,j_p) \end{cases} \tag{15}$$

where (i,j) represents the cell location in the respective matrix, the ith POS token in A and the jth POS token in B. S is a function (neither reflexive nor symmetric) that takes in two POS tokens and returns a score value. The extension gap penalty is represented by E and the opening gap penalty by F. Even with the newly created matrices and additional processing that must take place to populate the matrices, the running time will be $\mathcal{O}(n^2)$, with memory as $\mathcal{O}(n^2)$.

4 Results

4.1 Classification

A useful feature of OpBerg is its ability to determine whether a sentence is a good match to an existing labeled sentence; it thus captures the similarity between two sentences. This similarity measure then can be used to classify whether the sentence is causal. Furthermore, we seek the ability to classify a more diverse set of classes that are useful to biologists. We test the ability of OpBerg to classify causality against the baseline methods of logistic regression (LR), support vector machines (SVM), naive Bayes classifier (NBC), random forest (RF) [19], AGE, the local alignment algorithm (local), global alignment algorithm (global), k-means clustering, density-based spatial clustering of applications with noise (DBSCAN) [7], balanced iterative reducing and clustering using

Table 1. Datasets with their description, number of classes (C), documents (D) and sentences (S) used in the evaluation of OpBerg in the classification experiment.

Name	Description	C	D	S
RM46	ResearchMaps collection of 46 neuroscience articles	2, 4, 7	46	200
LLL05	Causal sentences and papers extracted from the LLL05 Challenge	2	45	131
NDE27	Non-domain experts labeling of various PubMed articles	2	27	1,025
BioCause	The biomedical discourse causality corpus and articles	2	20	1,000
RM6	Domain expert labeled set of articles from ResearchMaps	2, 4, 7	6	356

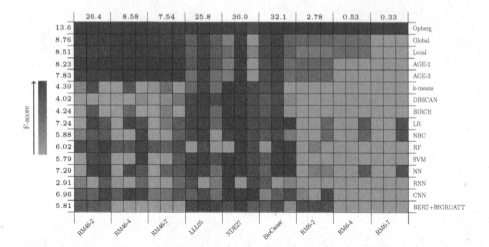

Fig. 1. F-scores shown as a heatmap for OpBerg compared with baseline methods using POS features of sentences, general word embeddings, and causal embeddings in order to determine causality and other biological classes. The sums of the total F-score for each dataset is given in the top most row. The total F-score for each model is displayed to the left of each heat map row.

hierarchies (BIRCH) [34], a feedforward neural network (NN), and a recurrent neural network (RNN), convolutional neural network (CNN) and lastly a BERT based bidirectional GRU with self attention (BERT+BIGRUATT) [18]. Other models were considered but are not shown due to brevity, not having significant differences between the shown baseline models. These other considered models were: term frequency-inverse document frequency (tf-idf), weakly supervised multilingual causality extraction from Wikipedia [12], Max-Matching and Attentive-Matching causal embedding models [2], and causal knowledge extraction through large-scale text mining [13]. Additionally each model was considered using a bag of words input as well as a combination of bag of words and parts of speech. These results are also omitted due to non significant results (compared to other baseline methods). All baseline models were implemented using their default parameters specified in their implementations. The various datasets used are described by Table 1.

Experimental Setup. For each dataset that contains only causal sentences we query the PubMed Central corpus for their respective articles. For each article we obtain the sentences which are not like a causal sentence. We define similar as the global alignment score and remove the top 5 most similar sentences to any causal sentence. All remaining sentences are labeled as non-causal. The resultant number of sentences and articles for each dataset is given in Table 1.

The hyperparameters for OpBerg were optimized using Bayesian optimization. For each labeled input set we then trained OpBerg, and the baseline methods using 10-fold cross validation. To obtain a classification probability for each alignment method, each test sentence's POS tokens (a) were compared to each POS-mapping of the sentences in the training set using the OpBerg algorithm. The highest matching score (H_m) was taken as the best match for a given sentence's POS string (b). If the best match belonged to a causal sentence, a probability was given to the test item as $P^+(a, H_m)$, with P^+ defined as:

$$P^+(a, H) = \frac{H}{Max[S(a_i, a_i)] \times |a|} \qquad (16)$$

For each test input, we also recorded the highest matching score aligned to a causal training sentence (H_c). If the best match was not a causal sentence, the probability was given to be:

$$P^-(a, H) = Min \left\{ \begin{array}{l} 1 - P^+(a, H_m) \\ P^+(a, H_c) \end{array} \right\} \qquad (17)$$

Inputs for all methods were made using part of speech (POS) features obtained by the Stanford CoreNLP POS tagger, embeddings using the GloVe Wikipedia 2014 + Gigaword dataset [26] and from a task-specific word embedding technique for causality [2].

In some of the datasets we were only able to obtain whether a sentence causal or not. However, in the ResearchMaps datasets we were able to determine more numerous classes of interest. The classes of interest were separated into three sets. The simplest set was labeled whether a sentence was causal. In the second, we are given the qualitative result of the experiment, i.e., increase, decrease, no change or non-causal. The most diverse set consisted of 7 different classes describing biological phenomena [29]: a permutation of the set negative or positive with the set of no change, increase, or decrease together with a non-causal label.

Experimental Results. The F-score for all models and datasets is represented as a heat map in Fig. 1. The intensity of each point is calculated with respect to other models in the same dataset. We normalize the points to emphasize the model differences when compared against the same data. To give a more global view of the discovered causality per dataset, we give the sums of the total F-score for each dataset in the top most row. To give an aggregate scoring of each model overall, the total F-score for each model is displayed to the left of each heat map row.

In all but one dataset, OpBerg gives the highest F-score amongst each baseline method. As we can see from Fig. 1, word embeddings do not improve the results much. For some models there was an improvement but overall there was not. We hypothesis this is because the nature of the text we are using. Many words which are important to causality are very esoteric words which may be difficult to obtain good vector representations for due to the lack of widespread use. We suppose a similar reasoning to why more causal-tailored methods [2,12,13,18] perform poorly at this task. It is likely OpBerg would not outperform these causal-tailored methods in general tasks of causality. However, in tasks in which esoteric terms and used, OpBerg may be well-suited to outperform other methods, due to the other methods reliance on general terminology for extracting causality (such as Wikipedia [12,26]).

Our results suggest alignments are a good technique in classifying causal sentences when the training set size is low; and the best among these alignment algorithms is OpBerg.

5   Discussion

Opberg represents a new approach to solving difficult NLP problems. Opberg is not meant to replace previous state of the art techniques in all datasets, rather it can be the best approach when the classification is complex and only a small amount (in the order of 10^2 to 10^4) labeled data exists. We maintain the effectiveness of Opberg over other machine learning approaches [3,4,8,11,14,20] when the labeled training data is small. If confidence is maintained in the quality of the knowledge source in covering the input data, predefined knowledge methods [6,17,25,27,33] may be more appropriate. In situations where an alternative method is preferred, Opberg can add value as an additional learning feature. In fact, it is the vision of the authors that Opberg will be most valuable to the research community as a complement to established approaches.

The benefits of Opberg come at an additional cost. The algorithm is complex, and the context can be unfamiliar to both the trained computer scientist and/or bioinformatician. One area that may be worth investigating is the use of graphics processing units (GPU) to reduce the execution time. If the GPU processing power can be used, large inputs sizes, such as the human genome, can be used as input to Opberg and results determined in a reasonable amount of time.

The approach of POS sequence alignment has some weaknesses. For one the algorithm does not take into account particles which may confer a different meaning than a sentence without the particle. An example of this would be the two sentences: *A does have a positive effect on B.* and *A does **not** have a positive effect on B.* Opberg does not take into account the negating word and a comparison would result in a high score. Another area for further research exploration is in entity extraction. Opberg may be good at finding similar structured sentences but in identifying the key terms, it is lacking. In real world applications, the authors envision to use Opberg in a pipeline with entity extraction methods run on the output after Opberg "filtering" for small labeled input sets.

6   Conclusion

This paper introduces a novel approach to causality discovery by considering alignments among POS mappings of sentences. This approach considers restrictions on the score size to break apart an alignment and enforces a minimum length requirement while also considering the number of alignments. OpBerg discovers meaningful alignments that are useful in finding semantic similarity of two causal sentences. The improved model and efficient implementation make OpBerg the best model to use when performing tasks that involve a small amount of labeled training data coupled together with esoteric text, particularly in that of causal extraction of biological research papers.

References

1. Abyzov, A., Gerstein, M.: Age: defining breakpoints of genomic structural variants at single-nucleotide resolution, through optimal alignments with gap excision. Bioinformatics **27**(5), 595–603 (2011)
2. Balashankar, A., Subramanian, L.: Learning faithful representations of causal graphs. In: ACL/IJCNLP 2021, 1–6 August 2021, pp. 839–850. Association for Computational Linguistics (2021)
3. Berger, M., Goldstein, E.: Increasing sentence-level comprehension through text classification of epistemic functions, pp. 139–150 (2021)
4. Blanco, E., Castell, N., Moldovan, D.I.: Causal relation extraction. In: Proceedings of the International Conference on Language Resources and Evaluation, LREC 2008, Marrakech, Morocco, 26 May–1 June 2008 (2008). http://www.lrec-conf.org/proceedings/lrec2008/summaries/87.html
5. Cao, P., et al.: Knowledge-enriched event causality identification via latent structure induction networks. In: ACL/IJCNLP 2021, Virtual Event, 1–6 August 2021, pp. 4862–4872. Association for Computational Linguistics (2021)
6. Cheng, Z., et al.: A unified target-oriented sequence-to-sequence model for emotion-cause pair extraction. IEEE/ACM Trans. Audio Speech Lang. Process. **29**, 2779–2791 (2021)
7. Ester, M., et al.: A density-based algorithm for discovering clusters in large spatial databases with noise. In: Proceedings of the Second International Conference on Knowledge Discovery and Data Mining (KDD-1996), Portland, Oregon, USA, pp. 226–231 (1996)
8. Fischbach, J., et al.: Fine-grained causality extraction from natural language requirements using recursive neural tensor networks. In: 29th IEEE International Requirements Engineering Conference Workshops, pp. 60–69. IEEE (2021)
9. Gong, J., Eldardiry, H.: Zero-shot relation classification from side information. In: Proceedings of the 30th ACM International Conference on Information & Knowledge Management, pp. 576–585 (2021)
10. Gu, Y., et al.: Domain-specific language model pretraining for biomedical natural language processing. ACM Trans. Comput. Healthc. (HEALTH) **3**, 1–23 (2021)
11. Han, R., et al.: ESTER: a machine reading comprehension dataset for reasoning about event semantic relations. In: EMNLP 2021, Virtual Event/Punta Cana, Dominican Republic, 7–11 November 2021, pp. 7543–7559. Association for Computational Linguistics (2021)
12. Hashimoto, C.: Weakly supervised multilingual causality extraction from Wikipedia. In: EMNLP/IJCNLP (1), pp. 2986–2997. Association for Computational Linguistics (2019)
13. Hassanzadeh, O., et al.: Causal knowledge extraction through large-scale text mining. In: AAAI, pp. 13610–13611. AAAI Press (2020)
14. Huang, F., Yates, A.: Open-domain semantic role labeling by modeling word spans. In: ACL 2010, pp. 968–978 (2010)
15. Huang, W., et al.: Local-to-global GCN with knowledge-aware representation for distantly supervised relation extraction. Knowl. Based Syst. **234**, 107565 (2021)
16. Huang, X., Chao, K.M.: A generalized global alignment algorithm. Bioinformatics **19**(2), 228–233 (2003)
17. Hussain, M., et al.: A practical approach towards causality mining in clinical text using active transfer learning. J. Biomed. Inform. **123**, 103932 (2021)
18. Kyriakakis, M., Androutsopoulos, I., Saudabayev, A., i Ametllé, J.G.: Transfer learning for causal sentence detection. In: BioNLP@ACL 2019, pp. 292–297 (2019)
19. Liaw, A., Wiener, M.: Classification and regression by randomForest. R News **2**(3), 18–22 (2002)

20. Lücking, A., Driller, C., Stoeckel, M., Abrami, G., Pachzelt, A., Mehler, A.: Multiple annotation for biodiversity: developing an annotation framework among biology, linguistics and text technology. Lang. Resour. Eval. 1–49 (2021). https://doi.org/10.1007/s10579-021-09553-5
21. Lyu, Q., et al.: Zero-shot event extraction via transfer learning: challenges and insights. In: ACL/IJCNLP 2021 (Volume 2: Short Papers), Virtual Event, 1–6 August 2021, pp. 322–332. Association for Computational Linguistics (2021)
22. Manning, C.D., et al.: The Stanford CoreNLP natural language processing toolkit. In: Association for Computational Linguistics (ACL) System Demonstrations, pp. 55–60 (2014)
23. Matiasz, N.J., et al.: Computer-aided experiment planning toward causal discovery in neuroscience. Front. Neuroinform. 11, 12 (2017)
24. Needleman, S.B., Wunsch, C.D.: A general method applicable to the search for similarities in the amino acid sequence of two proteins. J. Mol. Biol. 48(3), 443–453 (1970)
25. Patro, J., Baruah, S.: A simple three-step approach for the automatic detection of exaggerated statements in health science news, pp. 3293–3305 (2021)
26. Pennington, J., et al.: Glove: global vectors for word representation. In: A Meeting of SIGDAT, A Special Interest Group of the ACL, EMNLP 2014, Doha, Qatar, 25–29 October 2014, pp. 1532–1543. ACL (2014)
27. Saha, R., et al.: Using Tsetlin machine to discover interpretable rules in natural language processing applications. Expert Syst. e12873 (2021)
28. Shlain, M., et al.: Syntactic search by example. In: ACL 2020, 5–10 July 2020, pp. 17–23 (2020)
29. Silva, A.J., Müller, K.R.: The need for novel informatics tools for integrating and planning research in molecular and cellular cognition. Learn. Mem. 22(9), 494–498 (2015)
30. Smith, T.F., Waterman, M.S.: Identification of common molecular subsequences. J. Mol. Biol. 147(1), 195–197 (1981)
31. Taub-Tabib, H., et al.: Interactive extractive search over biomedical corpora. In: Proceedings of the 19th SIGBioMed Workshop on Biomedical Language Processing, BioNLP 2020, 9 July 2020, pp. 28–37 (2020). https://doi.org/10.18653/v1/2020.bionlp-1.3
32. Wada, T., et al.: Learning contextualised cross-lingual word embeddings and alignments for extremely low-resource languages using parallel corpora. In: Proceedings of the 1st Workshop on Multilingual Representation Learning, pp. 16–31 (2021)
33. Wang, Z., Wang, H., Luo, X., Gao, J.: Back to prior knowledge: joint event causality extraction via convolutional semantic infusion. In: Karlapalem, K., et al. (eds.) PAKDD 2021. LNCS (LNAI), vol. 12712, pp. 346–357. Springer, Cham (2021). https://doi.org/10.1007/978-3-030-75762-5_28
34. Zhang, T., et al.: BIRCH: an efficient data clustering method for very large databases. In: Proceedings of the 1996 ACM SIGMOD International Conference on Management of Data, pp. 103–114 (1996)
35. Zhang, Y., et al.: ReadsRE: retrieval-augmented distantly supervised relation extraction. In: SIGIR 2021, Virtual Event, Canada, 11–15 July 2021, pp. 2257–2262. ACM (2021)

Text-Based Causal Inference on Irony and Sarcasm Detection

Recep Firat Cekinel$^{(\boxtimes)}$ and Pinar Karagoz

Computer Engineering Department, Middle East Technical University, Ankara,
Turkey
{rfcekinel,karagoz}@ceng.metu.edu.tr

Abstract. The state-of-the-art NLP models' success advanced significantly as their complexity increased in recent years. However, these models tend to consider the statistical correlation between features which may lead to bias. Therefore, to build robust systems, causality should be considered while estimating the given task's data generating process. In this study, we explore text-based causal inference on the irony and sarcasm detection problem. Additionally, we model the latent confounders by performing unsupervised data analysis, particularly clustering and topic modeling. The obtained results also provide insight for the causal explainability in irony detection.

Keywords: Irony detection · Causal inference · Clustering · Topic modeling

1 Introduction

Traditional NLP models can achieve accurate prediction results using statistical correlations within data. However, performance of the conventional methods mainly depends on the data distribution of the training and testing datasets. For this reason, analyzing causal relationships which utilize the data generating process are helpful to create robust models [8,31]. More specifically, causal inference is a way of generating counterfactual explanations in hypothetical scenarios such as how the outcome variable is affected by an intervention on a treatment variable. The causal inference has been applied to create inferences on imaginary situations in several fields, but its practical applications in NLP have started to gain attention.

The cause-effect relationships of linguistic properties can be examined using causal inference by measuring the change in the outcome resulting from an intervention on a treatment. Under an imaginary scenario, the potential outcomes can be estimated by satisfying *ignorability*, *positivity*, and *consistency* assumptions (details given in Sect. 3.1). Usually, NLP applications rely on observational data, so randomly assigning texts is not feasible. In other words, to satisfy the ignorability assumption in observational studies while assigning treatment, there

R. Wrembel et al. (Eds.): DaWaK 2022, LNCS 13428, pp. 31–45, 2022.
https://doi.org/10.1007/978-3-031-12670-3_3

should not exist any unobserved confounders (predict both treatment and outcome). Identification is also another key aspect of causal inference for NLP, which suggests that the linguistic properties can be expressed using proxy labels [22,32,44]. Additionally, it is assumed that proxy labels can estimate the ground-truth causal relation of linguistic properties.

Many state-of-the-art NLP models can be considered black-box models, which receive text documents as input and generate an output dependent on the task. Therefore, explaining and intervening in the predictions of such models remain a challenging problem [5,8,35]. Some studies examined the applicability of causal methods to interpret the black-box NLP models by generating counterfactual statements [26]. These works can be classified as data perspective [38] and model component perspective [14,27] where the former is related to counterfactual statement generation and exploiting network artifacts is an example of the latter.

In this study, we focus on irony and sarcasm detection problem, and explore text-based causal inference by using the TextCause algorithm [32] to measure the causal effect of linguistic properties on this problem. Irony and sarcasm detection refers to way of verbal expressions such that the one's meaning is expressed through signifying just the opposite. Therefore, the problem includes difficulties and analyzing the causal relationships can provide insight for explainability of the generated models and improving the detection performance. The main contributions of this study can be summarized as follows:

- The causal effect of linguistic properties are examined in irony and sarcasm detection tasks using the TextCause algorithm [32].
- Latent confounders within text documents are modeled by using K-Means clustering and LDA topic modeling and their effects on the causal inference are analyzed.
- The obtained results provide insight in terms of the causal interpretability and explainability aspects.

2   Related Works

This study is built on top of the TextCause algorithm proposed by Pryzant et al. [32]. The authors use DistilBERT [39] language model to adjust text and they are inspired by Veitch et al.'s CausalBERT study [43] which adapts BERT to adjust texts as a confounder. Additionally, they generate causal embeddings using causal topic models, which were adopted from Blei et al. [1]. Keith et al. [17] summarize the methods to adjust texts for causal inference. Moreover, Fong et al. [11] discuss the required assumptions to use latent features of text as treatment. In another study, they also use topic modeling to discover latent treatments in texts [10]. Moreover, Wood-Doughty et al. [46] address the challenges of using proxy treatments for causal inference.

Recently, Yang et al. [47] conduct a survey of existing causality extraction methods for texts. Moreover, Feder et al. [8] provide a review of the use-cases of text-based causal inference and discuss fairness, interpretability, and robustness aspects. Texts can be considered as treatment [32,48], confounder [17,43], outcome [7] and even mediator [18] settings. Sridhar et al. [40] examine the causal effect of tone on online debates. Koroleva et al. [21] propose a model to measure the similarity of pairs of clinical trial outcomes and reports semantically using BERT-based language models.

There exist comprehensive studies that review models to explain black-box NLP models [5,8,26]. More recently, Chou et al. [4] also examine an in-depth review of the studies on model-agnostic counterfactual algorithms and argue that many such studies do not rely on causal theoretical formalism. Wang et al. [45] utilize a causal approach to exploit the attention weights of a sentiment classifier. Besides, perturbation-based approaches [23,35] have been used for explanation. Another prominent and challenging text-based causal explanation method is counterfactual statement generation [12,36,38] which requires manipulating text in a meaningful manner. Therefore, instead of modifying the text itself, changing its representation has emerged by [9,33]. Besides, Buyukbas et al. [2] work on the same Turkish tweet dataset as in this study and examine the explainability of transformer architectures using two popular explainability tools, LIME [35] and SHAP [23] for irony detection task. Likewise, Hazarika et al. [15] propose the CASCADE model that utilizes both contextual and content information to improve the sarcasm detection performance significantly on SARC [19] dataset.

3   Background

3.1   Causal Inference

Typical NLP models use statistical associations to make decisions and estimate the dataset's distribution using the training data. On the other hand, causal inference is an inverse problem that figures out the structural causal model of the data generating process, which leads to more robust and invariant models. Causal inference is about answering the counterfactual queries based on the intervention of interest. However, the counterfactual outcomes do not exist in the observational data in most cases. Therefore, the causal effect is the change of outcome variable Y by the intervention on treatment X when all other covariates are kept constant.

The initial step of causal inference represents the association between variables as Structural Causal Models (SCMs). The SCMs consist of directed acyclic graphs (DAGs) and a mathematical problem formulation. The variables are represented as nodes, and edges represent the causal relationship between variables.

Definition 1 (Structural Causal Model). *It consists of 3-tuples (U, V, E) where U denotes a set of exogenous variables (independent from the states), V denotes a set of endogenous variables (dependent to other states in the system) and they are connected by a set of structural equations, E, where each equation defines endogenous variables in terms of U and V.*

After representing the causal model as a graph, interventions on a treatment can be expressed using Pearl's do-calculus notation [30]. Three rules of do-calculus which allow to simulate interventions on treatment to identify causal relationships in DAGs are summarized below:

- *Rule 1:* Insertion and deletion of observations

$$P(Y \,|do(X), Z, W) = P(Y \,|do(X), Z), \text{ if W is irrelevant to Y}$$

- *Rule 2:* Action/observation exchange

$$P(Y \,|do(X), Z) = P(Y \,|X, Z), \text{ if Z blocks all back-door paths from X to Y}$$

- *Rule 3:* Insertion and deletion of actions

$$P(Y \,|do(X)) = P(Y), \text{ if there is no causal path from X to Y}$$

The first rule suggests that we can omit variables W if it is irrelevant to outcome Y. However, the second rule states that if variables Z blocks all backdoor paths from treatment X to Y, we must condition on Z. Finally, the third rule asserts that if there is no causal path from X to Y, we should not condition on X. A causal inference framework can estimate the counterfactual outcomes by making some assumptions that need to satisfy three criteria listed below:

- *Ignorability*: The treatment assignment and the counterfactual outcomes must be independent by randomizing the treatment assignment. However, for observational data, it is not feasible. Therefore, softer conditional ignorability criteria should be satisfied, which requires no unobserved confounders in the dataset.
- *Positivity*: For all covariates, the probability of receiving treatment must be greater than 0.
- *Consistency*: The outcome at unit i is only affected by the treatment at the same unit.

3.2 NLP with Causality

Texts are inherently high dimensional, and by encoding texts using language models, hidden factors such as topic, tone, and writing style can be discovered. BERT [6], a bi-directional transformer-based language model, had a breakthrough on NLP, which outperformed previous models on many tasks with significant margins. However, Feder et al. [8] indicated that such models utilize statistical relationships while making decisions. Therefore, their predictions can be considered unreliable. Moreover, McCoy et al. [25] pointed out that these language models may fail when the data distribution of the test set changes significantly since these models rely on some statistical reasonings. As a result, causal models are required to increase the models' generalization performance.

Secondly, the reasoning of any model can be evaluated with sensitivity and invariance tests. The former identifies how much minimal perturbation is necessary to switch the model's decision for the given sample. On the other hand, the latter determines whether a change in a causally unrelated feature impacts the model's decision. These tests can be valuable to interpret the model's robustness by feeding counterfactual inputs. Besides, Veitch et al. [42] stated that invariant models can perform better on different data distributions.

3.3   Causal Model Explainability

Language models such as BERT [6] are not inherently explainable. According to
Moraffah et al. [26] exploiting network artifacts such as attention weights is one
approach to infer the decisions of a neural model. However, these approaches can
only describe token-wise information. In addition, perturbating instances near
decision boundary [23,35] is another way of interpretability. Yet, sentence-level
estimates of such models may not be so successful [8]. In other words, these
approaches may result in erroneous explanations to the decision-makers since
they compute correlations between features [4,20,37].

In this context, causal models can generate counterfactual instances which
can be used for interpretability [8]. For instance, a data sample's prediction can
be compared with its counterfactual representative. More specifically, if a text
contains a concept, its counterfactual will not include that concept, and their
outputs can be compared to learn how the model makes decisions.

4   Methods

In this work, we investigate the causal inference for irony and sarcasm detection
problem, which involves text analysis. Therefore we apply text based causal
inference algorithm, TextCause, [32]. In addition to adapting TextCause to
irony/sarcasm detection problem, we extend the use of confounders by using
unsupervised data analysis.

4.1   Text-Based Causal Inference Using TextCause

TextCause, which is proposed by Pryzant et al. [32], employs the CausalBERT
model [43] that adjusts text for causal inference. The key innovation of the
TextCause algorithm is the assumption that neither the writer's intent nor the
reader's perception can be identified from observational data. Therefore, the
authors express the need to employ a proxy label $\hat{T}$ to estimate the causal effect
of a linguistic property. In other words, they train a proxy classifier to capture
both the writer's intent and the reader's perception. The proposed structural
causal model is presented in Fig. 1. According to this structural causal model, a
writer writes a text W that contains a linguistic property T with other covariates
Z. A reader's perception of that linguistic property is represented by $\widetilde{T}$ and $\widetilde{Z}$
and affects the outcome Y which can be estimated using a proxy label $\hat{T}$. Besides,
the authors state that the bias due to proxy treatment decreases as the proxy
classifier's accuracy increases. Therefore, for observational data, actual linguistic
property T can be measured using proxy labels $\hat{T}$.

The conditional ignorability assumption of causal inference requires that the
treatment assignment should be independent of the outcomes for observational
data. In other words, this assumption states that we need to adjust for all confounders to estimate the causal effect of the treatment. The causal effect can
be estimated using the Average Treatment Effect (ATE), which is formulated in
Eq. 1.

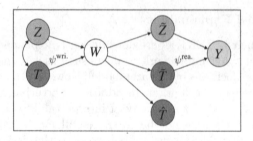

Fig. 1. The causal model in [32]

$$ATE = E[Y; do(T = 1)] - E[Y; do(T = 0)] \tag{1}$$

ATE can be expressed as the difference between the interventional outcome $(T=1)$ and the counterfactual outcome $(T=0)$. However, text documents may contain some hidden confounders, such as tone and writing style, so we need to adjust the ATE for all confounders using Pearl's backdoor-adjustment [29]. Since the authors use proxy labels to estimate the ATE, the modified ATE estimation is given in Eq. 2. The TextCause model uses DistilBERT to generate a representation of texts and employs the special classification token, CLS, to approximate the confounding information $\hat{Z}$. Therefore, the ATE estimator relies on the treatment, the language model representation of text and the one-hot encoding of the covariates. As a result, the model learns two vectors that corresponds to the language model representation and one-hot encoded covariates respectively.

$$ATE_{proxy} = E_W[E[Y|\hat{T} = 1, \widetilde{Z} = f(W)] - E[Y|\hat{T} = 0, \widetilde{Z} = f(W)]] \tag{2}$$

In addition to the text adjustment, another contribution of the TextCause algorithm is improving the recall of the proxy labels, which is motivated by lexicon induction [13] and label propagation [49]. The authors train logistic regression and pu-classifier models to predict proxy labels $\hat{T}^*$ and relabel the instances that labeled as $\hat{T} = 0$ but predicted as $\hat{T}^* = 1$. As a result, improved proxy labels and texts are required to measure the causal effect. Additional covariates and language model representation of a text should be adjusted as a confounder. Hence, the TextCause algorithm utilizes both proxy label improvement and text adjustment to estimate the causal effect of desired linguistic property.

4.2   Unsupervised Data Analysis for Determining Confounders

While applying text based causal inference on irony/sarcasm detection problem, the categories or groupings within the text collection is considered as a confounder. In order to determine the subgroups, two different techniques[1], topic modeling and clustering, are used.

---

[1] https://github.com/firatcekinel/Unsupervised-Data-Analysis.

Topic Modeling. Topic modeling is a statistical method to discover latent topics in a corpus. It is an unsupervised technique that examines semantic structures in a text. Moreover, the topics represent a group of similar words that are determined by statistical models. A document can be a mixture of several topics with different proportions based on a word's appearance in one of the topics. Therefore, a document can be classified using topic modeling based on the words' relevance to the abstract topics.

Latent Dirichlet Allocation (LDA) [1] is one of the most popular topic modeling techniques. It is a generative statistical model that uses the Dirichlet priors for word-topic and document-topic distributions and represents documents as a mixture of topics where the distribution over words determines the proportions. Given a corpus with M documents where a document w_i contains N-words and α and β are the Dirichlet prior parameters, the probability distribution of a document can be expressed as in Eq. 3. In this study, we lemmatized texts using SpaCy[2] and performed LDA to discover abstract topics that highlight several aspects of the document collection.

$$P(D, \alpha, \beta) = \prod_{m=1}^{M} \int P(\theta_m | \alpha)(\prod_{n=1}^{N} \sum_{Z_{mn}} P(Z_{mn} | \theta_m) P(W_{mn} | Z_{mn}, \beta)) m\theta_m \quad (3)$$

Clustering. Texts are inherently high-dimensional, so a text should be encoded to a latent vector space. Sentence embeddings map sentences to vectors that can measure semantic similarity between sentences or text summarization. Transformers [41] made a remarkable impact on NLP tasks that passed previous models with a substantial margin. Reimers et al. [34] introduce S-BERT, which is a transformer-based sentence embedding model. S-BERT was built on top of the pre-trained BERT [6] model but uses siamese and triplet networks to extract semantically meaningful sentence embeddings. The S-BERT produces large-sized vectors as sentence embedding, which should be transformed into a lower-dimensional space for clustering. Dimensionality reduction techniques such as PCA [16], and t-SNE [24] can be applied to transform high-dimensional data into a lower-dimensional space by preserving the meaningful information in the data.

Clustering is an unsupervised machine learning technique that groups similar data instances together. K-Means clustering is one of the most popular clustering methods that assign n data points to k clusters where each data point is assigned to a cluster whose cluster center is the nearest. Since unsupervised models do not have a ground truth, metrics such as the silhouette coefficient can measure the clustering quality. This study uses S-BERT to encode texts in a fixed-size latent space and applies dimensionality reduction using PCA or t-SNE. Finally, the transformed data is given to a K-Means model to group semantically similar texts.

---

[2] https://spacy.io/.

4.3   Modeling Causal Inference for Irony and Sarcasm Detection

In this work, we explore the cause-effect relationship for irony and sarcasm detection on two scenarios. The treatments (T), outputs (Y) and confounders (Z) considered in the scenarios are as follows.

Case 1. We measure the effect of writing sarcastic posts (T) on the popularity of the post, number of likes, (Y) and consider subreddit category, cluster label (by the K-Means model) and the topic category (by the LDA model) as confounder (Z), separately.

Case 2. We examine whether putting an exclamation mark (!) affects irony detection. In other words, we explore whether the exclamation mark (T) affects the readers' perception of a text as ironic (Y). The cluster label and topic category were also considered confounder (Z) in this scenario.

5   Experiments

5.1   Dataset and Settings

The first dataset that we use in our study is a Self-Annotated Reddit Corpus (SARC) [19] that contains 1.3 million sarcastic Reddit posts. It is a publicly-available dataset, and statements that end with "/s" marker, a common sarcastic marker of Reddit users, are annotated as sarcastic. Therefore, we can consider that the dataset might contain some false negative statements, such that there may be some statements that should be annotated as sarcastic but not marked as such. Moreover, we should not assume that all Reddit users know such markers, so the dataset might also contain some false positive statements. Secondly, we use a Turkish tweet dataset for irony detection [3, 28]. The dataset contains 300 non-ironic and 300 ironic tweets in Turkish, which were annotated manually.

The experiments are performed on Nvidia GeForce RTX 2080 Super GPU with 8 GB memory. The computer also includes Intel i7-8700k CPU@3.7 GHz with 12 cores. While implementing the model, Huggingface's multilingual DistilBERT [39] is used. It is a lighter BERT model that performs very close to the original model using significantly fewer parameters. Additionally, we performed some validation experiments to adjust hyperparameters such as epoch and learning rate. In Sect. 5.2, we present only the results with the best hyperparameter settings.

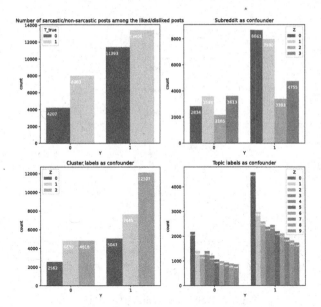

Fig. 2. Number of reddit posts for each confounder settings

5.2 Results

Case 1 Results. In this experiment, we assume that the subreddit category, topic label and cluster label affect the treatment and outcome, so we consider these attributes as confounder.

Firstly, we gather the posts in "AskReddit" $(Z=0)$, "news" $(Z=1)$, "world-news" $(Z=2)$, and "politics" $(Z=3)$ subreddits. If the posts' score is above five, we annotate them as "liked" comments. Besides if the posts' score is below 0, we annotate them as "disliked" comments. Overall, the number of comments satisfying these conditions are 37 K approximately. ,The number of popular (liked) posts within each confounder is given in Fig. 2.

Secondly, we assume that the LDA topic models could be used as a confounder. We measure the coherence score for various topic counts and observe that setting of 10 topics is a reasonable choice among a set of alternatives. The coherence score of this setting is 0.312. Likewise, we apply K-Means clustering to find optimal number of clusters with the collection of posts. According to Fig. 3, $K=3$ is sensible among the selected set of values according to elbow analysis. Additionally, for $K=3$, PCA and t-SNE plots are given in Fig. 4.

Finally, we measure the ATE score using the subreddit category, topic label, and cluster label as a confounder. Since the TextCause model requires proxy labels, we trained a BERT model using 400 K Reddit documents (80%–20% train-val sets) from other categories. The accuracy of the proxy classifier on the selected subreddits is 78.6%, and the f1-score is also calculated as 0.806. The TextCause model measures the oracle ATE value using the ground truth sarcastic label. The unadjusted ATE measures the treatment effect without adjusting for

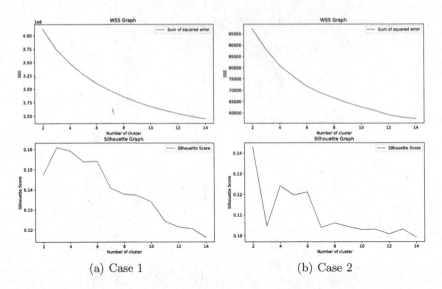

Fig. 3. WSS and silhouette plots

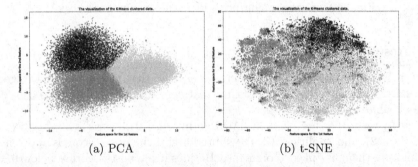

Fig. 4. K-Means clusters of reddit comments

any covariates. The T-boost values consider improved proxy treatments using pu classifier (to improve the recall for positive instances) and logistic regression. W adjust is another estimator that adjusts for text. Moreover, the last two estimates combine W adjust with T-boost.

We trained the TextCause algorithm for five epochs. According to the ATE scores in Table 1, adjusting for the topic label, cluster label, and subreddit category improves the ATE result. The oracle value suggests that the sarcastic writing style increases the chance of a post being liked between 6% and 10%. Additionally, the closest estimations are predicted by the T-boost reg model, and the TextCause models' subreddit and cluster label estimations are very close to the oracle estimator. However, when we adjust for topic labels, the unadjusted ATE estimator, which calculates ATE without adjusting for any covariate, becomes the second closest estimator overall.

Table 1. Case 1: Subreddit, topic and cluster labels were considered as confounder

Estimator	$ATE_{subreddit}$	ATE_{lda}	$ATE_{k-means}$
Oracle	0.0773	0.1029	0.0669
Unadjusted	0.1041	0.1041	0.1041
T-boost reg	**0.0742**	**0.1037**	**0.0639**
T-boost pu	0.0670	0.1005	0.0549
W adjust	0.0644	0.0659	0.0725
TextCause pu	0.0676	0.0776	0.0635
TextCause reg	0.0735	0.0719	0.0746

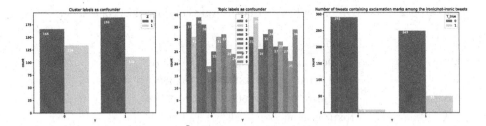

Fig. 5. Number of tweets for each confounder settings

Case 2 Results. In this experiment, we measure the effect of using an exclamation mark (!) on the irony. Since the treatment is evident, there is no need for a proxy label. We evaluate the causal question on the Turkish irony dataset, which is annotated by [3,28]. As in the first experiment, we consider the topic and cluster labels as a confounder. Figure 5 indicates the number of tweets for each confounder settings. According to the WSS and silhouette plots given in Fig. 3, the highest silhouette score is measured when $K = 2$. The clusters projected with the PCA and t-SNE are presented in Fig. 6. On the other hand, for LDA model, 10 topics settings is a reasonable choice since the coherence score of this setting is measured as 0.7318.

We trained the TextCause algorithm for 15 epochs. According to the ATE results that are presented in Table 2, the treatment has a considerable impact on the posts' irony. However, contrary to our expectations, there is an inverse relationship between the treatment and the outcome. As seen in Fig. 5, this is possibly due to that the number of ironic tweets that contain an exclamation mark is just 17% (51 out of 300 tweets) of the all ironic tweets. In addition, text adjustment for LDA topic labels estimates the closest prediction to the oracle value. However, for cluster labels the unadjusted setting was the closest among the all estimators. Note that, we do not present the results of the T-boost estimators because proxy labels were not appropriate in this setting.

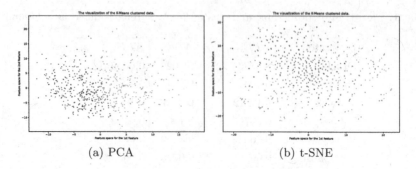

(a) PCA   (b) t-SNE

Fig. 6. K-Means clusters of tweets

Table 2. Case 2: Topic and cluster labels were considered as confounder

Estimator	$ATE_{k-means}$	ATE_{lda}
Oracle	−0.3955	−0.3451
Unadjusted	**−0.3889**	−0.3889
W adjust	−0.3506	**−0.3383**
TextCause pu	−0.0581	−0.0570
TextCause reg	−0.0292	−0.0204

6   Conclusion

This study addresses the application of causal inference to text analysis. Specifically, we employ the TextCause algorithm [32] to estimate the causal effect of sarcastic linguistic properties on a text's popularity, and use of punctuations, particularly (!) on understanding/detecting irony. Moreover, we perform unsupervised data analysis using clustering and topic modeling and utilize these methods' output for the causal inference. According to the measurements, cluster and topic labels may contain latent information on ironic linguistic properties and the popularity of the posts. The results can be reexamined in-depth in terms of explainability for future work. For instance, counterfactual statements that do not contain a specific linguistic property can be generated and fed into the causal-text model. The results can be examined in terms of invariance and sensitivity.

References

1. Blei, D.M., Ng, A.Y., Jordan, M.I.: Latent dirichlet allocation. J. Mach. Learn. Res. **3**(Jan), 993–1022 (2003)
2. Buyukbas, E.B., Dogan, A.H., Ozturk, A.U., Karagoz, P.: Explainability in Irony detection. In: Golfarelli, M., Wrembel, R., Kotsis, G., Tjoa, A.M., Khalil, I. (eds.) DaWaK 2021. LNCS, vol. 12925, pp. 152–157. Springer, Cham (2021). https://doi.org/10.1007/978-3-030-86534-4_14

3. Cemek, Y., Cidecio, C., Öztürk, A.U., Çekinel, R.F., Karagöz, P.: Investigating the neural models for Irony detection on Turkish informal texts. In: 2020 28th Signal Processing and Communications Applications Conference (SIU), pp. 1–4. IEEE (2020)
4. Chou, Y.L., Moreira, C., Bruza, P., Ouyang, C., Jorge, J.: Counterfactuals and causability in explainable artificial intelligence: theory, algorithms, and applications. Inf. Fusion **81**, 59–83 (2022)
5. Danilevsky, M., Qian, K., Aharonov, R., Katsis, Y., Kawas, B., Sen, P.: A survey of the state of explainable AI for natural language processing. In: Proceedings of the 1st Conference of the Asia-Pacific Chapter of the Association for Computational Linguistics and the 10th International Joint Conference on Natural Language Processing, pp. 447–459 (2020)
6. Devlin, J., Chang, M.W., Lee, K., Toutanova, K.: BERT: pre-training of deep bidirectional transformers for language understanding. arXiv preprint arXiv:1810.04805 (2018)
7. Egami, N., Fong, C.J., Grimmer, J., Roberts, M.E., Stewart, B.M.: How to make causal inferences using texts. arXiv preprint arXiv:1802.02163 (2018)
8. Feder, A., et al.: Causal inference in natural language processing: estimation, prediction, interpretation and beyond. arXiv preprint arXiv:2109.00725 (2021)
9. Feder, A., Oved, N., Shalit, U., Reichart, R.: Causalm: causal model explanation through counterfactual language models. Comput. Linguist. **47**(2), 333–386 (2021)
10. Fong, C., Grimmer, J.: Discovery of treatments from text corpora. In: Proceedings of the 54th Annual Meeting of the Association for Computational Linguistics (Volume 1: Long Papers), pp. 1600–1609 (2016)
11. Fong, C., Grimmer, J.: Causal inference with latent treatments. Am. J. Polit. Sci. (2019)
12. Gardner, M., et al.: Evaluating models' local decision boundaries via contrast sets. In: Findings of the Association for Computational Linguistics: EMNLP 2020, pp. 1307–1323 (2020)
13. Hamilton, W.L., Clark, K., Leskovec, J., Jurafsky, D.: Inducing domain-specific sentiment lexicons from unlabeled corpora. In: Proceedings of the Conference on Empirical Methods in Natural Language Processing. Conference on Empirical Methods in Natural Language Processin, vol. 2016, p. 595. NIH Public Access (2016)
14. Harradon, M., Druce, J., Ruttenberg, B.: Causal learning and explanation of deep neural networks via autoencoded activations. arXiv preprint arXiv:1802.00541 (2018)
15. Hazarika, D., Poria, S., Gorantla, S., Cambria, E., Zimmermann, R., Mihalcea, R.: Cascade: contextual sarcasm detection in online discussion forums. In: Proceedings of the 27th International Conference on Computational Linguistics, pp. 1837–1848. Association for Computational Linguistics (2018). http://aclweb.org/anthology/C18-1156
16. Jolliffe, I.: Principal Component Analysis. Springer, Cham (2002). https://doi.org/10.1007/b98835
17. Keith, K., Jensen, D., O'Connor, B.: Text and causal inference: a review of using text to remove confounding from causal estimates. In: Proceedings of the 58th Annual Meeting of the Association for Computational Linguistics, pp. 5332–5344 (2020)
18. Keith, K., Rice, D., O'Connor, B.: Text as causal mediators: research design for causal estimates of differential treatment of social groups via language aspects. In: Proceedings of the First Workshop on Causal Inference and NLP, pp. 21–32 (2021)

19. Khodak, M., Saunshi, N., Vodrahalli, K.: A large self-annotated corpus for sarcasm. In: Proceedings of the 11th International Conference on Language Resources and Evaluation (LREC 2018) (2018)
20. Kilbertus, N., Rojas Carulla, M., Parascandolo, G., Hardt, M., Janzing, D., Schölkopf, B.: Avoiding discrimination through causal reasoning. In: Advances in Neural Information Processing Systems, vol. 30 (2017)
21. Koroleva, A., Kamath, S., Paroubek, P.: Measuring semantic similarity of clinical trial outcomes using deep pre-trained language representations. J. Biomed. Inform. **100**, 100058 (2019)
22. Lucy, L., Demszky, D., Bromley, P., Jurafsky, D.: Content analysis of textbooks via natural language processing: findings on gender, race, and ethnicity in Texas US history textbooks. AERA Open **6**(3), 2332858420940312 (2020)
23. Lundberg, S.M., Lee, S.I.: A unified approach to interpreting model predictions. In: Advances in Neural Information Processing Systems, vol. 30 (2017)
24. Van der Maaten, L., Hinton, G.: Visualizing data using t-SNE. J. Mach. Learn. Res. **9**(11) (2008)
25. McCoy, T., Pavlick, E., Linzen, T.: Right for the wrong reasons: diagnosing syntactic heuristics in natural language inference. In: Proceedings of the 57th Annual Meeting of the Association for Computational Linguistics, pp. 3428–3448 (2019)
26. Moraffah, R., Karami, M., Guo, R., Raglin, A., Liu, H.: Causal interpretability for machine learning-problems, methods and evaluation. ACM SIGKDD Explor. Newsl. **22**(1), 18–33 (2020)
27. Narendra, T., Sankaran, A., Vijaykeerthy, D., Mani, S.: Explaining deep learning models using causal inference. arXiv preprint arXiv:1811.04376 (2018)
28. Ozturk, A.U., Cemek, Y., Karagoz, P.: IronyTR: Irony detection in Turkish informal texts. Int. J. Intell. Inf. Technol. (IJIIT) **17**(4), 1–18 (2021)
29. Pearl, J.: Causality. Cambridge University Press, Cambridge (2009)
30. Pearl, J.: The do-calculus revisited. In: Proceedings of the 28th Conference on Uncertainty in Artificial Intelligence, pp. 3–11 (2012)
31. Pearl, J., Mackenzie, D.: The Book of Why: The New Science of Cause and Effect. Basic Books, New York (2018)
32. Pryzant, R., Card, D., Jurafsky, D., Veitch, V., Sridhar, D.: Causal effects of linguistic properties. In: Proceedings of the 2021 Conference of the North American Chapter of the Association for Computational Linguistics: Human Language Technologies, pp. 4095–4109 (2021)
33. Ravfogel, S., Prasad, G., Linzen, T., Goldberg, Y.: Counterfactual interventions reveal the causal effect of relative clause representations on agreement prediction. In: Proceedings of the 25th Conference on Computational Natural Language Learning, pp. 194–209 (2021)
34. Reimers, N., Gurevych, I.: Sentence-BERT: sentence embeddings using siamese BERT-networks. In: Proceedings of the 2019 Conference on Empirical Methods in Natural Language Processing and the 9th International Joint Conference on Natural Language Processing (EMNLP-IJCNLP), pp. 3982–3992 (2019)
35. Ribeiro, M.T., Singh, S., Guestrin, C.: Model-agnostic interpretability of machine learning. arXiv preprint arXiv:1606.05386 (2016)
36. Ribeiro, M.T., Wu, T., Guestrin, C., Singh, S.: Beyond accuracy: behavioral testing of NLP models with checklist. In: Proceedings of the 58th Annual Meeting of the Association for Computational Linguistics, pp. 4902–4912 (2020)
37. Richens, J.G., Lee, C.M., Johri, S.: Improving the accuracy of medical diagnosis with causal machine learning. Nat. Commun. **11**(1), 1–9 (2020)

38. Ross, A., Wu, T., Peng, H., Peters, M.E., Gardner, M.: Tailor: generating and perturbing text with semantic controls. arXiv preprint arXiv:2107.07150 (2021)
39. Sanh, V., Debut, L., Chaumond, J., Wolf, T.: DistilBERT, a distilled version of BERT: smaller, faster, cheaper and lighter. arXiv preprint arXiv:1910.01108 (2019)
40. Sridhar, D., Getoor, L.: Estimating causal effects of tone in online debates. In: International Joint Conference on Artificial Intelligence (2019)
41. Vaswani, A., et al.: Attention is all you need. In: Advances in Neural Information Processing Systems, vol. 30 (2017)
42. Veitch, V., D'Amour, A., Yadlowsky, S., Eisenstein, J.: Counterfactual invariance to spurious correlations: why and how to pass stress tests. arXiv preprint arXiv:2106.00545 (2021)
43. Veitch, V., Sridhar, D., Blei, D.: Adapting text embeddings for causal inference. In: Conference on Uncertainty in Artificial Intelligence, pp. 919–928. PMLR (2020)
44. Voigt, R., et al.: Language from police body camera footage shows racial disparities in officer respect. Proc. Natl. Acad. Sci. **114**(25), 6521–6526 (2017)
45. Wang, Y., Huang, M., Zhu, X., Zhao, L.: Attention-based LSTM for aspect-level sentiment classification. In: Proceedings of the 2016 Conference on Empirical Methods in Natural Language Processing, pp. 606–615 (2016)
46. Wood-Doughty, Z., Shpitser, I., Dredze, M.: Challenges of using text classifiers for causal inference. In: Proceedings of the Conference on Empirical Methods in Natural Language Processing. Conference on Empirical Methods in Natural Language Processing, vol. 2018, p. 4586. NIH Public Access (2018)
47. Yang, J., Han, S.C., Poon, J.: A survey on extraction of causal relations from natural language text. Knowl. Inf. Syst. **64**, 1161–1186 (2022)
48. Zhang, J., Mullainathan, S., Danescu-Niculescu-Mizil, C.: Quantifying the causal effects of conversational tendencies. Proc. ACM Hum. Comput. Interact. **4**(CSCW2), 1–24 (2020)
49. Zhu, X., Ghahramani, Z.: Learning from labeled and unlabeled data with label propagation (2002)

Sarcastic RoBERTa: A RoBERTa-Based Deep Neural Network Detecting Sarcasm on Twitter

Maciej Hercog[1], Piotr Jaroński[1], Jan Kolanowski[1,3], Paweł Mieczyński[1],
Dawid Wiśniewski[1(✉)] (ID), and Jedrzej Potoniec[1,2] (ID)

[1] Institute of Computing Science, Poznan University of Technology,
ul. Piotrowo 2, 60-965 Poznań, Poland
Dawid.Wisniewski@cs.put.poznan.pl
[2] Center for Artificial Intelligence and Machine Learning, Poznan University
of Technology, ul. Piotrowo 2, 60-965 Poznań, Poland
[3] Cisco Systems Poland, Warszawa, Poland

Abstract. Sarcastic RoBERTa is an approach to recognizing sarcastic
tweets written in English. It is based on a pre-trained RoBERTa model
supported by a 3-layer feed-forward fully-connected neural network. It
establishes a new state-of-the-art result on the *iSarcasm* dataset, attaining the F_1 score of 0.526, and being not far from the human performance
of 0.616.

Keywords: Text classification · Sarcasm detection · RoBERTa ·
iSarcasm · SARC · Knowledge transfer

1 Introduction

*Sarcasm is the use of remarks that clearly mean the opposite of what they say,
made in order to hurt someone's feelings or to criticize something in a humorous
way* [1]. Recognizing it purely from a short piece of text, such as a post on social
media, is an interesting research challenge. First, the notion itself is challenging,
not only for an algorithm but also for a human, as it requires disregarding the
literal meaning of a text in favor of the opposite meaning. Moreover, short pieces
of text lack context. In some cases, they may be, e.g., a response to some other
post offering some context, but frequently they rely on an implicit context known
both to the author and the intended readers, e.g., related to the current sports
events. Additionally, an utterance intended as sarcastic on one day may be read
as non-sarcastic only a few days later when some new facts become available.
Finally, as sarcasm purposefully twists and turns the meaning of language, its
form and intended understanding may vary from author to author to a much
greater extent than for straightforward utterances.

This work summarizes the results of a bachelor thesis by PM, PJ, JK, MH, done under
the supervision of DW and JP.

R. Wrembel et al. (Eds.): DaWaK 2022, LNCS 13428, pp. 46–52, 2022.
https://doi.org/10.1007/978-3-031-12670-3_4

In this work, we focus on detecting sarcasm in posts from Twitter[1] – a popular social media platform. Such posts are called *tweets* and are currently up to 280 characters long, including special-purpose parts, such as mentioning other users with @ marker or tagging content with #. We leverage two datasets, *iSarcasm* [17] and the Self-Annotated Reddit Corpus (*SARC*) [13]. We describe them in Sect. 2. We introduce *Sarcastic RoBERTa*, a deep neural network based on a pre-trained deep neural language model RoBERTa [15], and describe its details in Sect. 3, along with simpler models used for an ablation study. Sarcastic RoBERTa establishes a new state-of-the-art result on the *iSarcasm*'s test set, attaining the F_1 score equal to 0.526, whereas human annotators scored 0.616. The details of the evaluation are presented in Sect. 4. We discuss the related work in Sect. 5, and conclude in Sect. 6.

The main contributions of the paper are as follows:

- we show that transformer models, in particular RoBERTa, are viable tools for sarcasm detection;
- we present a training procedure leading to Sarcastic RoBERTa, a model improving over the current state of the art by 0.062 on the F_1 score;
- we publish the model for ease of reuse.

2 Datasets

In this paper, we employ two datasets: *iSarcasm*[2] [17] and the Self-Annotated Reddit Corpus (*SARC*) [13]. *iSarcasm* is a dataset consisting of tweets IDs, each assigned with a binary label indicating whether the tweet is sarcastic or not. Additionally, each sarcastic tweet is assigned to one of the following categories of sarcasm: *sarcasm, irony, satire, understatement, overstatement, rhetorical question*. The tweets were collected in a survey, where each participant was supposed to link one sarcastic tweet of theirs and three non-sarcastic tweets. The dataset is heavily biased in some aspects, as 97% of participants were from the US and the UK, and over 72% were younger than 35 years old. The final assignment of labels and categories was done by trained linguists, yielding 777 sarcastic and 3,707 non-sarcastic tweets. The authors split the dataset into a training set and a test set in the proportion of 80:20. *iSarcasm* reveals that even the problem of deciding whether a tweet is sarcastic or not is not an easy one. In an experiment where three third-party annotators labeled the test set, with their votes aggregated by the majority voting, they scored only 0.616 on the F_1 score.

In this work, we disregard the sarcasm category and concentrate only on deciding whether a tweet is sarcastic or not. We employed the Twitter API to download the textual content of the tweets. As some of the tweets are no longer available, the final dataset consists of 3535 tweets, i.e., 78% of the original

---

[1] www.twitter.com.

[2] https://github.com/silviu-oprea/iSarcasm.

dataset, divided into a training set of 2825 tweets (480 sarcastic) and a test set of 710 tweets (119 sarcastic).

The other dataset, *SARC*, is a large corpus derived automatically from Reddit[3], a social media platform for interest-focused communities. *SARC* consists of 533 M comments, out of which 1.34 M (i.e., 0.2%) are sarcastic. A comment was labeled as sarcastic if it terminates with the marker /s, which is an established way on Reddit to self-label sarcastic utterances. In an experiment using 100 pairs consisting of a sarcastic and a non-sarcastic comment with the goal of deciding which one is sarcastic, the majority voting over 5 human annotators attained the accuracy of 92%. In this work, we use a balanced subset of *SARC* consisting of about 1.01 M comments, available as the file `train-balanced-sarcasm.csv` at https://www.kaggle.com/danofer/sarcasm.

3   Proposed Approach

Architecture. To construct Sarcastic RoBERTa, a binary classifier capable of distinguishing between sarcastic and non-sarcastic tweets, we used an approach inspired by [18]. The input of the classifier is a string of characters preprocessed in such a way that every token starting with @ is replaced with a general `@user` placeholder. Similarly, every URL is replaced with `http` token.

First, the input is passed through a tokenizer appropriate for a given RoBERTa model that was obtained using the `AutoTokenizer` class provided in the HuggingFace's `transformers` library. This step transforms any sequence into tokens known by RoBERTa. Each sequence is then padded to a length of 127 tokens.

Then, we use a Twitter pre-trained RoBERTa model [15]. The model is available in the HuggingFace repository[4] as `cardiffnlp/twitter-roberta-base`. It was trained on a corpus of 58M tweets and exhibited a non-negligible performance improvement over the general purpose `roberta-base` model when compared on the *TweetEval* benchmark [5].

We use the last hidden state of RoBERTa as an embedding representing the input, and we apply to it a 1D-max-pooling layer with a kernel size of 20. Next, we use two fully-connected feed-forward layers (resp. 800 and 20 neurons), each followed by a dropout layer ($p = 0.5$) and the ReLU activation function. Finally, the output layer consists of 2 neurons with the softmax activation function.

Training Protocol. We employed the cross-entropy loss and the Adam optimizer [14] with a learning rate set to 10^{-5}. First, we used randomly selected 400,000 examples from the balanced subset of *SARC* and pre-trained our model over a single epoch using mini-batches of size 30. We then reset the optimizer and trained the model over 80% of the training set of *iSarcasm* dataset for 15 epochs using mini-batches of size 40. After each epoch, we saved the model along with its F_1 score on the validation set (the remaining 20%), and we selected the model

---

[3] https://www.reddit.com/.
[4] https://huggingface.co/.

with the highest F_1 score. It was the model after two epochs of training, and it is available for download at https://bit.ly/3JpQWiF.

Ablation Study. To evaluate the influence of pre-training, we constructed two alternative models, following the same architecture and training procedure, except: (1) for the model denoted *without* SARC, *with Twitter pre-training* we did not pre-train on the *SARC* dataset; (2) for the model denoted *without* SARC, *without Twitter pre-training* we did not pre-train on the *SARC* dataset, and used the `roberta-base` model instead of the one pre-trained on Twitter.

4 Evaluation

We use F_1 score, precision, and recall calculated on the *iSarcasm* test set to evaluate our model and its variants without pre-training. We report the scores in Table 1. We also report the results of a baseline classifier constructed using *fastText* [12]. For comparison purposes, we also report the best results from the recent works of Guo et al. [9], Handoyo et al. [10], and the human score.

Table 1. Evaluation scores

Variant	F_1	Precision	Recall
Sarcastic RoBERTa	0.526	0.507	0.546
without *SARC*, with Twitter pre-training	0.485	0.475	0.495
without *SARC*, without Twitter pre-training	0.401	0.408	0.394
fastText	0.258	–	–
LOANT [9]	0.464	0.436	0.497
Handoyo et al. [10]	0.404	–	–
Human annotation [17]	0.616	0.550	0.701

Sarcastic RoBERTa, employing pre-training on both the corpus of 58M tweets, and the *SARC* dataset outperforms the current state-of-the-art approaches by a fair margin. This underscores the importance of pre-training and exhibits how powerful transfer learning is.

5 Related Work

Social media platforms are an interesting and important source of raw data for researchers almost since their start. In particular, the problem of recognizing sarcastic tweets was first posed by Davidov et al. in 2010 [6], who collected a dataset of tweets and used distant supervision to label them according to the presence of tags such as #sarcasm. Numerous attempts similar in spirit were undertaken over the years, e.g., by Barbieri et al. [4], Ptácek et al. [19], or Bamman and Smith [3].

Another approach to labeling was to do it manually, either by experts or volunteers (e.g., Abercrombie and Hovy [2]). Labeling using crowdsourcing, albeit prevalent in other domains (e.g., Filatova used it to detect sarcasm in Amazon reviews [7]), seems to be non-existent in this particular area. Some authors used a combination of both approaches, where tweets were first filtered based on used hashtags, and then the labels were manually verified (e.g., in Task 3 of SemEval-2018 by Van Hee et al. [23]).

Initially, manually constructed features and simple classifiers were used in the task, e.g., Davidov et al. used pattern mining and k-nearest neighbors (k-NN). While these classical approaches are still used (e.g., in recent work by Sundararajan and Palanisamy [22]), current solutions based on neural language models are becoming more and more popular. For example, Ren et al. [20] used a model based on a convolutional neural network (CNN), whereas Gregory et al. [8] used recurrent neural networks and transformers.

Very recently Moores and Mago [16] published a survey about automatic sarcasm detection on Twitter, offering a comprehensive view on the topic.

Besides sarcasm detection, there are numerous other similar tasks related to recognizing non-verbal and contextual aspects of utterances. For example, Janiszewski et al. [11] considered the problem of conversation breakdown in social media, whereas Singh and Toshniwal [21] predicted the next geo-spatial location of a user based on their tweets.

6   Discussion and Conclusions

In this work, we presented Sarcastic RoBERTa, a deep neural model for detecting sarcastic tweets in English. The proposed solution, based on the RoBERTa model, outperforms the current state of the art on the *iSarcasm* dataset by a fair margin. While the attained F_1 score of 0.526 may seem low at the first glance, one must consider that whether some utterance is sarcastic depends on its discourse context (e.g., a sentence sarcastic in one discourse is not necessarily sarcastic in another), time context (e.g., an utterance may cease to be sarcastic when new information is revealed), it differs from person to person (e.g., what one sees as sarcastic, may not be sarcastic for another), and it differs between the author of an utterance and its recipients (intended vs perceived sarcasm).

Due to some tweets being no longer available, our work only included a subset of the original *iSarcasm* dataset. While we have no control over which tweets disappeared, and thus believe this to be a random subset, the resulting dataset may be simpler than the original one. Deciding whether this is the case would require using one of the earlier approaches on the used subset, which was deemed out of scope for this work.

The results of the ablation study presented in Sect. 4 underline how much one can gain from transfer learning. First, simply replacing the basic RoBERTa model with a model trained on tweets increased the F_1 score by 0.084 by priming the model for the general language of tweets. Then, introducing additional pre-training on *SARC* increased the score by another 0.041 by priming the model to the phrases characteristic to sarcasm. Unfortunately, such pre-training is costly, and we were unable to investigate the results of using, e.g., the whole *SARC* corpus, or other sarcasm-related datasets.

Acknowledgements. This work was supported by the Statutory Funds of the Poznan University of Technology.

References

1. Sarcasm. In: McIntosh, C. (ed.) Cambridge Advanced Learner's Dictionary, 4th edn. Cambridge University Press (2013)
2. Abercrombie, G., Hovy, D.: Putting sarcasm detection into context: the effects of class imbalance and manual labelling on supervised machine classification of Twitter conversations. In: Proceedings of the ACL 2016 Student Research Workshop, pp. 107–113 (2016)
3. Bamman, D., Smith, N.A.: Contextualized sarcasm detection on Twitter. In: Cha, M., et al. (eds.) ICWSM 2015, pp. 574–577. AAAI Press (2015)
4. Barbieri, F., et al.: Modelling sarcasm in Twitter, a novel approach. In: Balahur, A., et al. (eds.) WASSA@ACL 2014, pp. 50–58. ACL (2014)
5. Barbieri, F., et al.: TweetEval: unified benchmark and comparative evaluation for tweet classification. In: Cohn, T., et al. (eds.) EMNLP 2020, pp. 1644–1650. ACL (2020)
6. Davidov, D., et al.: Semi-supervised recognition of sarcasm in Twitter and Amazon. In: Lapata, M., Sarkar, A. (eds.) CoNLL 2010, pp. 107–116. ACL (2010)
7. Filatova, E.: Irony and sarcasm: corpus generation and analysis using crowdsourcing. In: Calzolari, N., et al. (eds.) LREC 2012, pp. 392–398. ELRA (2012)
8. Gregory, H., et al.: A transformer approach to contextual sarcasm detection in Twitter. In: Proceedings of the 2nd Workshop on Figurative Language Processing, pp. 270–275. ACL (2020)
9. Guo, X., et al.: Latent-optimized adversarial neural transfer for sarcasm detection. In: NAACL-HLT 2021, pp. 5394–5407. ACL (2021)
10. Handoyo, A.T., et al.: Sarcasm detection in Twitter - performance impact while using data augmentation: word embeddings. CoRR abs/2108.09924 (2021)
11. Janiszewski, P., Lango, M., Stefanowski, J.: Time aspect in making an actionable prediction of a conversation breakdown. In: Dong, Y., Kourtellis, N., Hammer, B., Lozano, J.A. (eds.) ECML PKDD 2021. LNCS (LNAI), vol. 12979, pp. 351–364. Springer, Cham (2021). https://doi.org/10.1007/978-3-030-86517-7_22
12. Joulin, A., Grave, E., Bojanowski, P., Mikolov, T.: Bag of tricks for efficient text classification. arXiv preprint arXiv:1607.01759 (2016)
13. Khodak, M., et al.: A large self-annotated corpus for sarcasm. In: Calzolari, N., et al. (eds.) LREC 2018. ELRA (2018)
14. Kingma, D.P., Ba, J.: Adam: a method for stochastic optimization. In: Bengio, Y., LeCun, Y. (eds.) ICLR 2015 (2015)
15. Liu, Y., et al.: RoBERTa: a robustly optimized BERT pretraining approach. CoRR abs/1907.11692 (2019)
16. Moores, B., Mago, V.: A survey on automated sarcasm detection on Twitter. CoRR abs/2202.02516 (2022)

17. Oprea, S., Magdy, W.: iSarcasm: a dataset of intended sarcasm. In: ACL 2020 (2020)
18. Potamias, R.A., Siolas, G., Stafylopatis, A.G.: A transformer-based approach to irony and sarcasm detection. Neural Comput. Appl. **32**(23), 17309–17320 (2020). https://doi.org/10.1007/s00521-020-05102-3
19. Ptácek, T., et al.: Sarcasm detection on Czech and English Twitter. In: Hajic, J., Tsujii, J. (eds.) COLING 2014, pp. 213–223. ACL (2014)
20. Ren, Y., et al.: Context-augmented convolutional neural networks for Twitter sarcasm detection. Neurocomputing **308**, 1–7 (2018)
21. Singh, R., Toshniwal, D.: Location prediction using sentiments of Twitter users. In: Ordonez, C., Bellatreche, L. (eds.) DaWaK 2018. LNCS, vol. 11031, pp. 98–108. Springer, Cham (2018). https://doi.org/10.1007/978-3-319-98539-8_8
22. Sundararajan, K.T., Palanisamy, A.: Multi-rule based ensemble feature selection model for sarcasm type detection in Twitter. Comput. Intell. Neurosci. **2020**, 2860479:1–2860479:17 (2020)
23. Van Hee, C., et al.: SemEval-2018 task 3: irony detection in English tweets. In: SemEval 2018, pp. 39–50. ACL (2018)

A Fast NDFA-Based Approach to Approximate Pattern-Matching for Plagiarism Detection in Blockchain-Driven NFTs

Darius Galiș[1]([✉])[iD], Ciprian Pungilă[1][iD], and Viorel Negru[1,2][iD]

[1] West University of Timișoara, Timișoara, Romania
{darius.galis,ciprian.pungila,viorel.negru}@e-uvt.ro
[2] ICAM - Environmental Advanced Research Institute, Timișoara, Romania

Abstract. We are presenting a fast and innovative approach to performing approximate pattern-matching for plagiarism detection, using an NDFA-based approach that significantly enhances performance compared to other existing similarity measures. We outline the advantages of our approach in the context of blockchain-based non-fungible tokens (NFTs). After testing in real-world scenarios, we conclude that our approach is suitable and adequate to perform approximate pattern-matching for plagiarism detection, yet significantly faster and therefore more suitable for big data analysis.

Keywords: Plagiarism detection · Pattern-matching · Approximate · Blockchain · NFT · Automaton · Aho-Corasick · Sliding window · Similarity measurement

1 Introduction

Plagiarism is an important issue with respect to protecting intellectual property, a crucial centerpiece in the academic community, as well as in various business aspects. Plagiarism occurs whenever material is copied without the author's permission or approval. With the spread of the Internet, plagiarism becomes a growing concern to authors developing original work. The blockchain industry aims to resolve this problem through the use of non-fungible tokens (NFTs), a market that has grown exponentially in the past few years [1,2].

2 Related Work

Numerous papers and approaches have been proposed to leverage plagiarism detection, and improve existing techniques to identify potential scammers. For example, in [3], the authors propose an approach based on tokenization and the computation of a similarity score, making use of natural language processing

R. Wrembel et al. (Eds.): DaWaK 2022, LNCS 13428, pp. 53–58, 2022.
https://doi.org/10.1007/978-3-031-12670-3_5

(NLP) techniques. In [4], authors focus on a hybrid approach: by combining the Jaccard and cosine distances, they recurse to machine learning (ML) and NLP, corroborated with text mining and similarity analysis. Similar to the previous approach, the authors recurse to tokenization. Saini et al. [5] had build a plagiarism detection using text mining methods and the cosine distance for computing similarity. Putri et al. [6] propose a text-mining-based approach to plagiarism detection using a Karp-Rabin variant of pattern-matching, involving a slightly modified Jaccard similarity computation method and tokenization, to identify the number of common words in the two texts being compared.

There are several classes and types of plagiarism. From a lexical perspective, text mining is a common approach to performing plagiarism detection. From a semantic perspective, as outlined before, there are approaches based on semantics, particularly NLP and ML, that bring the concept of "understanding" of the text contents into the picture, just as well. Our main focus in this paper is the approach based on text mining utilizing the tokenization of input data, in order to perform the mining process and assign a plagiarism degree between two different texts, and in particular to protecting intellectual property stored through the form of an NFT that was already previously minted in the blockchain.

3   Implementation

Our approach is inspired by the Aho-Corasick [7] finite state machine, however it differs significantly from it as it creates a new NDFA state machine, using a sliding window concept at node-level, as well as a locally-applied similarity measurement for the sliding window concept that we introduce. We modify the failure transitions of the NDFA, and we are changing the entire heuristics of those transitions so that we can apply inexact pattern-matching. In order to achieve this, we allow the transitions from a certain state to have multiple outcomes when parsing the same input characters. In order to start determining the output of any given transition, we employed the usage of the sliding window concept, accompanied by the computation of every possible suffix for every node in our automaton.

Our proposed NDFA also adds a new pre-processing step, computed after the failure functions calculation inside of the automaton. Here, for every node in our automaton, we will start parsing every possible transition, or when that is not possible, the failure transition linked to that node, so that in the end, we compute all the viable suffixes (of a length equal to that of a fixed-length sliding window) that emerge from that particular node. This approach effectively produces, for every node, a unique list of all possible matches from that particular node forward, significantly reducing matching comparisons when the local optimum is employed. Preprocessing complexity is $O(W * L * S_L)$ (W - number of keywords, L - average length of a keyword, and S_L - the sliding window length).

The nondeterministic behavior is properly highlighted in the actual processing stage. At this stage, the novelty brought by this NDFA approach is the possibility of applying local similarity measurements for approximate pattern-matching at any given step inside the parsing of the automaton for the length of

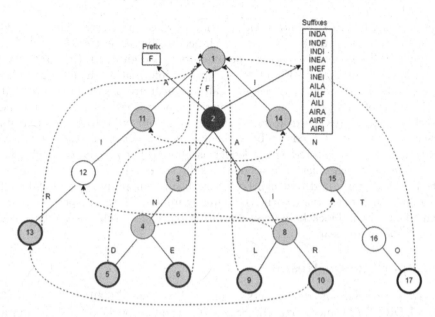

Fig. 1. The non-deterministic multiple pattern matching automaton for the set of keywords {*AND, FIND, FINE, FAIL, FAIR, INTO*}, using a sliding window of length 4. Parsed nodes used for computing suffixes are colour marked. (Color figure online)

the sliding window, instead of computing them at every position inside the input, for the length of the entire keyword at the processing stage (Fig. 1). We chose to focus on the Euclid, Hamming and Levenshtein distances, as the similarity measures used for the local optimum threshold computation process.

By obtaining a locally determined result of the similarity metric applied at any given node (between a concatenation of the node's prefix and the suffixes, one at a time), and the string of the same length resulting by parsing the input to get to the respective node, we can decide if we move throughout the automaton with an existing transition (similar to the exact match of the Aho-Corasick DFA), or we jump to the mismatch transition, thus allowing multiple outcome states for the same input. The distinction between the aforementioned cases is done by an *a priori* definition of a local optimum threshold value for all the similarity metrics results. We hypothesize that choosing a high value for this local optimum threshold may result in certain keywords not being detected (false negatives), while setting the value too low, might produce other matches than expected (false positives). In order to try to mitigate the occurrence of false positives, we also added an additional verification so no match is detected if more than a given empirical percentage of the length is disjoint. The percentage is set based on the maximum length of the keywords and the sliding windows. For example, for a maximum keyword length of 10 characters, the existing similarity measurements for approximate pattern-matching are using a percentage of 20%, while for the same keyword length size and a sliding window of 5, in order to have related similarities, the percentage of our approach need to be set as 10%.

While NFTs may have become popular due to digital art (in particular, photographs), they may also hold other types of information, including books, various texts, novels, short stories, etc., all of which are meant to ensure that specific form of digital art can be later traded or otherwise preserved in the blockchain. In our proposed approach, in order for an NFT to be minted, its associated digital data needs to be verified against plagiarism. After it passes verification, it can be stored in the ledger as minted. For text-based NFTs, we propose an approach where the original text is stored alongside the tokenized text. This way, tokenized text can be verified using our proposed NDFA-based approach against plagiarism, with other already-minted NFTs. As speed is a concerning factor in all blockchain-drive platforms, our focus on this aspect becomes stronger and more clear.

4   Experimental Results

We have performed our experiments on an AMD Ryzen 5 5600H CPU, with 16 GB of DDR4 RAM and a 512 GB SSD drive, running on Windows 10, and a C programming language implementation of NFT text plagiarism detection in our custom blockchain. For testing, we have used slightly modified versions of the datasets created by Paul Clough (Information Studies) and Mark Stevenson (Computer Science), at the University of Sheffield [8,9]. For scenario 1, we benchmarked the performance of our approach using a pair of datasets of 75 patterns to be found (of variable lengths, ranging between 4 to 15 characters), inside of an input size of 2,450 KB (or in terms of book pages, approximately 2,000 book pages and 376 NDFA nodes in total). For scenario 2, we used the same number of patterns, but we doubled the lengths of the original scenario 1, obtaining a total of 1,875 NDFA nodes. We also ran consecutive tests in order to empirically determine the best local optimum threshold values, so that the results produced by our approach match the ones produced by the classic similarity measurements used: Euclid, Hamming and Levenshtein.

Figure 2 showcases how the outcome is influenced by changing the local optimum threshold values. We correctly presumed that setting the value too high will cause false negatives. Also, due to the increased number of false positives, when the local optimum threshold is set too low, we would be dealing with an additional set of false negatives. This is caused by partial matches both in the sliding window and the input data, which produce better local similarity scores as opposed to the global score, associated with the entire pattern. Therefore, we recommend that the sliding window length is at least as high as the length of the longest pattern in our set of keywords. In terms of accuracy, the number of false negatives in Fig. 2 is a constant 1 achieved for the Euclidean similarity, throughout the entire experiment. Both Hamming and Levensthein similarities return no false negatives when the threshold value is set to 0.75, which seems to be the best fit for the scenarios tested. The number of false positives remains constant for the Euclidean similarity, but follows a decreasing trend as the threshold

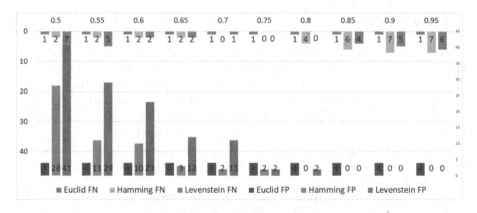

Fig. 2. Number of false negatives (FN) and false positives (FP), for various local threshold values (0.5 to 0.95).

values get closer to 1 for the others (which, in theory, would produce the same exact matching results as in the original Aho-Corasick DFA) - particularly, starting with a threshold value of 0.85, both Hamming and Levenshtein similarities produce no false positives.

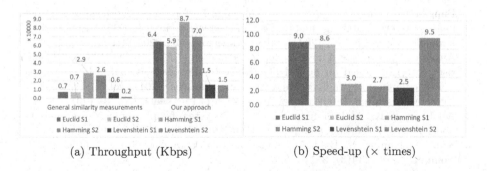

(a) Throughput (Kbps)   (b) Speed-up (× times)

Fig. 3. Performance comparison between classic similarity measures and our proposed approach in both scenarios (S1 and S2)

For thorough testing, we have empirically observed that the longer the patterns in the keyword set tend to be, the faster the speed-up achieved for the Levenshtein similarity. The results for both scenarios are being outlined in Fig. 3. We have observed significant run-time speed-ups for our proposed approach, compared to the classic approaches. In particular, for the dataset tested in scenario 2, the Levenshtein similarity measurement throughput has improved to about 9.5× higher throughput, up from 2.5× higher throughput as in scenario 1. The Euclid-driven similarity measurement has achieved a speed-up of about 9× in both scenarios tested, and similarly the Hamming distance provided around 3× higher throughput in our approach.

5   Conclusions

In this paper, we have presented an innovative fast approach to performing approximate-pattern matching for plagiarism detection, with particular applicability to blockchain-driven, NFT-ready platforms and ecosystems. For our proposed implementation, we have used our own NDFA-based automaton along with a sliding window concept and local thresholds at node-level, for tracing partial matches faster. We have tested our approach and concluded that it behaves suitably similar to existing various other similarity measures used in text mining for plagiarism detection, while obtaining a significant speed-up improvement of up to 9.5× faster throughput.

This research was supported by the virtuaLedger project [10] and the MOISE project number 240/2020, ID POC/398/1/1, financed by EU, Romanian government and West University of Timisoara. The views expressed in this paper do not necessarily reflect those of the corresponding project's partners.

References

1. Howcroft, E.: Reuters. NFT sales hit $25 billion in 2021, but growth shows signs of slowing. https://www.reuters.com/markets/europe/nft-sales-hit-25-billion-2021-growth-shows-signs-slowing-2022-01-10/. Accessed 2 Apr 2022
2. Pungila, C., Galis, D., Negru, V.: A New High-Performance Approach to Approximate Pattern-Matching for Plagiarism Detection in Blockchain-Based Non-Fungible Tokens (NFTs) (2022). https://arxiv.org/abs/2205.14492
3. Anzelmi, D., Carlone, D., Rizzello, F., Thomsen, R., Hussain, D.M.A.: Plagiarism detection based on SCAM algorithm. In: Proceedings of the International MultiConference on Engineers and Computer Scientists 2011, vol. I, pp. 272–277. Newswood Limited, International Association of Engineers, IAENG (2011)
4. AL-Jibory, F.K., Al-Tamimi, M.S.H.: Hybrid system for plagiarism detection on a scientific paper. Turk. J. Comput. Math. Educ. **12**(13), 5707–5719 (2021)
5. Saini, A., Kumari, S., Bahl, A., Singh, M.: Plagiarism checker: text mining. Int. J. Comput. Appl. **134**(3), 8–11 (2016)
6. Putri, R.E., Putera, A., Siahaan, U.: Examination of document similarity using Rabin-Karp algorithm. Int. J. Recent Trends Eng. Res. **3**(8), 196–201 (2017). https://doi.org/10.23883/IJRTER.2017.3404.4SNDK
7. Aho, A.V., Corasick, M.J.: Efficient string matching: an aid to bibliographic search. Commun. ACM **18**(6), 333–340 (1975). https://doi.org/10.1145/360825.360855
8. Clough, P., Stevenson, M.: Developing a corpus of plagiarised short answers. Lang. Resour. Eval. **45**(1), 5–24 (2011). https://doi.org/10.1007/s10579-009-9112-1
9. Clough, P., Stevenson, M.: Developing a corpus of plagiarised short answers (2011). https://s3.amazonaws.com/video.udacity-data.com/topher/2019/January/5c4147f9_data/data.zip. Accessed 2 Apr 2022
10. The virtuaLedger project. https://virtualedger.com. Accessed 2 Apr 2022

Data Warehousing and OLAP

On Decisive Skyline Queries

Akrivi Vlachou[1]($\boxtimes$), Christos Doulkeridis[2]($\boxtimes$), João B. Rocha-Junior[3],
and Kjetil Nørvåg[4]

[1] University of the Aegean, Samos, Greece
`avlachou@aegean.gr`
[2] University of Piraeus, Piraeus, Greece
`cdoulk@unipi.gr`
[3] Universidade Estadual de Feira de Santana, Feira de Santana, Brazil
`joao@uefs.br`
[4] Norwegian University of Science and Technology, Trondheim, Norway
`noervaag@ntnu.no`

Abstract. Skyline queries aim to identify a set of interesting objects that balance different user-specified criteria, i.e., that have values as good as possible in *all* specified criteria. However, objects whose values are good in only a subset of the given criteria are also included in the skyline set, even though they may take arbitrarily bad values in the remaining criteria. To alleviate this shortcoming, we study the decisive subspaces that express the semantics of skyline points and determine skyline membership. We propose a novel query, called *decisive skyline query*, which retrieves a set of points that balance all specified criteria. Our experimental study shows that the newly proposed query is more informative for the user.

Keywords: Skyline query · Decisive subspaces · Decisive skyline query

1 Introduction

Skyline queries [1,5,8] constitute a powerful tool for data analysis and multi-objective optimization, as they enable balancing of different (and often conflicting) criteria specified by the user. Such queries return a set of data points (*skyline points*) that are not dominated by any other point in all dimensions. A point p *dominates* another point q, if p is better than or equal to q in all dimensions and strictly better than q in at least one dimension. Nevertheless, the skyline set contains also points that fail to balance among all given criteria, as we demonstrate in the following.

Example 1. (Motivating example) Assume that a tourist is interested in booking a hotel with a low price, a good ranking based on the customers' ratings, and nearby the beach. To this end, the tourist performs skyline analysis using an online hotel database in order to discover hotels that fulfill all criteria. In Fig. 1, the hotels belonging to the skyline set of the hotel database are depicted. However, by inspecting this result set, we observe that several hotels have good values

© The Author(s), under exclusive license to Springer Nature Switzerland AG 2022
R. Wrembel et al. (Eds.): DaWaK 2022, LNCS 13428, pp. 61–73, 2022.
https://doi.org/10.1007/978-3-031-12670-3_6

	Price	Distance	Rank
Panorama Hotel	$70	600m	8
Heaven Hotel	$50	100m	16
Beach Hotel	$60	1000m	4
Imperial Resort	$200	200m	6
City Center Hotel	$160	1500m	1
Sunset Hotel	$20	300m	20
Paradise Hotel	$10	1800m	18

Fig. 1. Hotels in the skyline of a hotel database.

on subsets of the available criteria only, but not all of them. For instance, *City Center Hotel* is included in the result due to its rank (minimum value), despite the fact that it has the second worst value in distance. Similarly, *Paradise Hotel* has the lowest price, but it fails to balance the remaining two criteria. The same holds for *Sunset Hotel*, which has the worst rank among the skyline points and is included in the result set because of its combined values in distance and price. However, had the user been really interested in such a hotel, she would have specified as criteria only a subset of the dimensions, namely distance and price. In this paper, we argue that the user needs a new query type that excludes from its results set hotels that may have arbitrarily bad values. In this example, only *Panorama Hotel* would satisfy the constraint imposed by this new query. We call this new query type *decisive skyline query*.

As shown in the example, the skyline query always returns the data point with the best value in one dimension, regardless of the values in the other dimensions, as this point cannot be dominated by any other point. Put differently, the skyline definition imposes "OR semantics" between the different criteria. In the hotel database, the skyline set contains the hotels that are the best trade-offs among (a) rank, price and distance, OR (b) rank, price, OR (c) price, distance, OR have (d) the minimum price, OR (e) the minimum rank, OR (f) the minimum distance. But this is not the objective of the user's search, since the user is looking for the best trade-offs among rank, price and distance.

An indirect consequence of the aforementioned "OR semantics" is that the skyline cardinality [4,7,21] increases rapidly with the dimensionality of the data space. The high cardinality of the skyline set originates from the fact that as the number of criteria increases, the combinations of different criteria increase exponentially. In turn, the probability that a point is dominated in all different combinations decreases, thus leading to more skyline points. Intuitively, it is more difficult to satisfy more criteria, therefore it would be expected that with increasing the number of criteria, the result size should decrease (or stay constant). Should we add too many criteria, none of the points will be able to satisfy all of them, thus resulting into an empty result set. In contrast to this intuition, the cardinality of skyline set increases with increasing dimensionality.

Most existing approaches focus on the effect of the problem and try to restrict the skyline cardinality, often motivated by the controlled output size of top-k queries [6,13,19]. Towards this goal, different categories of approaches have been recently proposed, including (1) selecting k representative skyline points [10,12,16,17], (2) restricting the skyline cardinality by changing the dominance relationship [2], and (3) ranking the skyline points based on different metrics [3,11,18] or user-defined functions [9].

In this paper, we take a radically different approach. We address what we consider to be the cause of the problem, and not the effect. To this end, we focus on the semantics of skyline queries (first studied by Pei et al. [15]). Informally, the *decisive subspaces* of a skyline point are responsible for the point being part of the skyline set, i.e., its values in these dimensions qualify it as skyline point. Capitalizing on this concept, we propose a novel query type, called *decisive skyline query*. We investigate two variants of the decisive skyline query, the *strict* variant, which returns only the subset of skyline points that have the full space as decisive subspace, and the *relaxed* variant, which returns also points with decisive subspaces that *cover* the entire data space. Interestingly, as a by-product, it turns out that the decisive skyline query does not suffer from increased output size for increased dimensionality. We emphasize that this is the first paper that focuses on the significance of retrieving points based on the properties of their decisive subspaces, since in [15] the aim was to find the subspace skyline points of all subspaces.

The rest of the paper is organized as follows. In Sect. 2, we present the background knowledge for our approach, the necessary definitions and formally state the problem. In Sect. 3, we present useful pruning properties for efficient computation of the decisive skyline query, and we describe a novel index-based algorithm. Section 4 follows presenting the results of the experimental evaluation. Finally, we conclude the paper and sketch future research directions in Sect. 5.

2   Problem Formulation

Given a data set P on a data space $\mathcal{D}$ defined by a set of m dimensions $\{d_1, \ldots, d_m\}$, a data object $p \in P$ is represented as an m-dimensional point $p = \{p[1], \ldots, p[m]\}$ where $p[i]$ is the value on dimension d_i. A point $p \in P$ dominates another point $q \in P$, denoted as $p \prec q$, if (1) on every dimension d_i, $p[i] \leq q[i]$; and (2) on at least one dimension d_j, $p[j] < q[j]$. The *skyline* $\mathcal{S}(P)$ is a set of points which are not dominated by any other point in P. Without loss of generality, we assume that skylines are computed with respect to min conditions on all dimensions and that all values are non-negative.

The notion of skyline can be extended to subspaces. Each non-empty subset U of $\mathcal{D}$ ($U \subseteq \mathcal{D}$) is referred to as a *subspace* of $\mathcal{D}$. The *skyline* of a subspace $U \subseteq \mathcal{D}$ is a set $\mathcal{S}_U(P) \subseteq P$ which are not dominated by any other point on subspace U. As shown in [15,20], the skyline set of the full space does not contain all the subspace skyline points of the different subspaces. A skyline point q in $\mathcal{S}_U(P)$ is either a skyline point in $\mathcal{S}_V(P)$ (assuming $U \subset V$) or there exists another data point p, such that $p[i] = q[i]$ ($\forall d_i \in U$), that dominates q on the dimension set $V - U$.

2.1   Intuition of Decisive Subspaces

Let us first assume that the distinct value condition holds, which means that no two points share the same value in a given dimension (i.e., for any two points p and q of P it holds that $\forall d_i \in \mathcal{D} : p[i] \neq q[i]$). In this case, any subspace skyline point also belongs to the skyline set of the full space, which in turn simplifies the definition of the decisive skyline queries. Under the distinct value condition, the *decisive subspace* [15] of a skyline point p is defined as follows.

Definition 1. *(Decisive subspace) For a skyline point $p \in \mathcal{S}(P)$, a subspace U of $\mathcal{D}$ is called decisive, if (1) p is a subspace skyline in U ($p \in \mathcal{S}_U(P)$), and (2) there exists no subspace $V \subset U$ such that p is a subspace skyline point in V ($\nexists V \subset U$ such that $p \in \mathcal{S}_V(P)$).*

A skyline point p can have multiple decisive subspaces. We use $DecSub(p)$ to denote the set of decisive subspaces for a skyline point p. If a point p has a decisive subspace $U \subset \mathcal{D}$, then this fact alone promotes p to become a full space skyline, irrespective of p's values in dimension set $\mathcal{D} - U$. Obviously, such skyline points may not balance the remaining dimensions.

  To address the problems of the semantics of the traditional skyline operator, we define the *strict decisive skyline* set $\mathcal{DS}(P)$ as the set of skyline points that have the full space $\mathcal{D}$ as their decisive subspace, i.e., $\mathcal{DS}(P) = \{p | p \in \mathcal{S}(P)$ and $DecSub(p) = \mathcal{D}\}$. Based on the definition of decisive subspaces for the case of distinct values, a skyline point p belongs to the decisive skyline set, if there does not exist any other subspace $U \subset \mathcal{D}$ for which p belongs to the subspace skyline set ($\nexists U \subset \mathcal{D}$ such that $p \in \mathcal{S}_U(P)$). We argue that decisive skyline points are guaranteed to have good values in all given criteria, in contrast to subspace skyline points.

  The above definition imposes the semantics of decisive skyline sets in a strict (or rigid) way. A more relaxed variant, denoted $\widehat{\mathcal{DS}}(P)$, is also defined as follows: $\widehat{\mathcal{DS}}(P) = \{p | p \in \mathcal{S}(P)$ and $\bigcup_{(\forall U_i \in DecSub(p))} U_i = \mathcal{D}\}$. This *relaxed decisive skyline* set also includes points that belong to subspace skyline sets, as long as their decisive subspaces *cover* the full space. Thus, the relaxed decisive skyline points also balance all criteria, but possibly in different subspaces that cover the full space. Also, notice that by definition $\mathcal{DS}(P) \subseteq \widehat{\mathcal{DS}}(P)$.

Example 2. Consider a data space $\mathcal{D} = ABC$ and a data set P defined in $\mathcal{D}$ (Fig. 2(a)). All points are skyline points and Fig. 2(b) depicts their decisive subspaces. For p_7, subspace A is the decisive subspace, therefore the value of A is sufficient to qualify p_7 as a skyline point in the full space independently of its values in the other dimensions. Similar for p_2 and p_5 the decisive subspaces are B and C respectively. On the other hand, AC is also a decisive subspace for p_2, because p_2 appears in the subspace skyline of AC, and AC is not a super-set of B. Only point p_3 has the full space ABC as decisive subspace, and this is the only point in this example that belongs to the decisive skyline set $\mathcal{DS}(P) = \{p_3\}$. Points p_1 and p_2 may also be considered as good options, even though they do not belong to $\mathcal{DS}(P)$. For example, p_2 has the best value in dimension B, but

	A	B	C		DecSub( )
p_1	4	5	2	p_1	{AC, AB}
p_2	3	1	6	p_2	{B, AC}
p_3	5	4	4	p_3	{ABC}
p_4	7	2	3	p_4	{AC}
p_5	6	6	1	p_5	{C}
p_6	2	3	8	p_6	{AB}
p_7	1	7	7	p_7	{A}

(a) Data set   (b) *DecSub*

Fig. 2. Example of decisive subspaces.

also balances nicely dimensions AC, since it is in the subspace skyline in AC. Points p_1 and p_2, together with p_3, belong to the relaxed decisive skyline set $\widehat{\mathcal{DS}}(P) = \{p_1, p_2, p_3\}$.

2.2 Formal Definition of Decisive Subspaces

In the following, we withdraw the restriction on points taking distinct values. In the general case, the main difference is that there may exist subspace skylines that do not belong to the full space skyline points. Recall that a subspace skyline point $q \in \mathcal{S}_U(P)$ is either a skyline point in the full space or there exists another data point p, such that $p[i] = q[i]$ ($\forall d_i \in U$), that dominates q on the dimension set $\mathcal{D} - U$. If such a point p exists, then the remaining $\mathcal{D} - U$ dimensions are important to determine whether q qualifies as a skyline point.

Definition 2. *(Maximal set of non-distinct points) Given a set of points G and set of dimensions U, we define as maximal set of non-distinct points the set:* $\mathcal{O}(G, U) = \{p_i | p_i \in P, \ \forall d_k \in U \ and \ \forall p_j \in G : p_i[k] = p_j[k]\}$.

Based on the above definition, $\mathcal{O}(G, U)$ is the maximal set of points of P with identical values with the points of G in U, i.e., there exists no other point $q \in P$ with this property. The following definition is equivalent to the definition of [15].

Definition 3. *(Maximal skyline group) Given a set of points G and set of dimensions U, the pair $\{G, U\}$ is called maximal skyline group and is denoted as $SG(G, U)$, if it holds that (1) $\forall p_i \in G$ it holds that $p_i \in \mathcal{S}_U(P)$ (2) $\forall p_i, p_j \in G$ and $\forall d_k \in U$ $p_i[k] = p_j[k]$ (3) $\nexists d_k \in \mathcal{D} - U$ such that $\forall p_i, p_j \in G : p_i[k] = p_j[k]$ (4) $\nexists p_j \in P - G$ such that $\exists p_i \in G$ and $\forall d_k \in U : p_i[k] = p_j[k]$.*

Intuitively, $SG(G, U)$ is the maximal set of points with same values in U, these points are subspace skylines in U, and U is the maximal set of dimensions for which this set of points coincide. In the following, we define the concept of decisive subspaces for a maximal skyline group.

Definition 4. *(Decisive subspaces of maximal skyline group) Given a maximal skyline group $SG(G,U)$, a subspace $V \subseteq U$ is called decisive for $SG(G,U)$ if (1) $\forall p_i \in G$ it holds that $p_i \in \mathcal{S}_V(P)$ (2) $\mathcal{O}(G,V) = G$ (3) $\nexists V' \subset V$ such that conditions 1) and 2) hold for V'.*

2.3   Decisive Skyline Points

We denote the decisive subspaces of the maximal skyline group $SG(G,\mathcal{D})$ as $DecSub(G)$. A decisive subspace V of $SG(G,U)$ means that all points in G share the same values in U and are in the subspace skyline set for every subspace V' such that $V \subseteq V' \subseteq U$. We cannot conclude if all points of G belong to the skyline set of $\mathcal{D}$, since depending on the remaining dimensions some of them may be dominated. Only if they are incomparable in the remaining dimensions, then all of them will belong to the skyline set. Skyline points that belong to a maximal skyline group $SG(G,\mathcal{D})$ that has the full space as a decisive subspace are included to the skyline set based on the values of all given dimensions, regardless if these points form groups in some subspaces.

Definition 5. *(Strict decisive skyline points) A skyline point p belongs to the strict decisive skyline set $\mathcal{DS}(P) \subseteq \mathcal{S}(P)$ of a data set P, if there exists a maximal skyline group $SG(G,\mathcal{D})$ such that $p \in G$ and the decisive subspace of G is the full space $(DecSub(G) = \{\mathcal{D}\})$.*

Definition 6. *(Relaxed decisive skyline points) A skyline point p belongs to the relaxed decisive skyline set $\widehat{\mathcal{DS}}(P) \subseteq \mathcal{S}(P)$ of a data set P, if there exists a maximal skyline group $SG(G,\mathcal{D})$ such that $p \in G$ and the union of the decisive subspaces of G is the full space $\left(\bigcup_{\forall U_i \in DecSub(G)} U_i = \{\mathcal{D}\}\right)$.*

	A	B	C
P_1	4	8	5
P_2	1	6	10
P_3	10	2	1
P_4	1	10	1

(a) Data set

$S_{ABC}(P)=\{p_1,p_2,p_3,p_4\}$
$S_{AB}(P)=\{p_2,p_3\}$
$S_{BC}(P)=\{p_3\}$
$S_{AC}(P)=\{p_4\}$
$S_A(P)=\{p_2,p_4\}$
$S_B(P)=\{p_3\}$
$S_C(P)=\{p_3,p_4\}$

(b) $S_U(P)$

	DecSub()
p_1	{ABC}
p_2	{AB}
p_3	{B}
p_4	{AC}
p_2, p_4	{A}
p_3, p_4	{C}

(c) DecSub

Fig. 3. Example of decisive skyline set.

Example 3. Consider the data set P depicted in Fig. 3(a). The decisive subspaces for each maximal skyline group $SG(G,U)$ are shown in Fig. 3(c). We observe that point p_1 is the only point that belongs to the decisive skyline set. Point p_3 has B as decisive subspace. In turn, this means that p_3 belongs to the skyline set independently of its values in the other dimensions. In this example, p_3 has the worst value of all points in dimension A. On the other hand, point p_2 has AB as decisive subspace and not A, even though it is subspace skyline in A. This is because it coincides with p_4 in A and they form a group $\{p_2, p_4\}$ in that subspace. Still, the value of p_2 in dimension C does not influence whether p_2 belongs to the skyline set or not, thus this value can be arbitrarily high. Note that in this small example, the strict and relaxed decisive skyline points are the same.

3 Decisive Skyline Algorithm

A straightforward way to compute the decisive skyline set is to compute all maximum skyline groups and their decisive subspaces. Then, the points that belong to a group that have the full space as a decisive subspace can be easily determined. Computing all maximum skyline groups requires evaluating all $2^m - 1$ subspace skyline queries and requires multiple disk accesses on the same data. We refer to this approach as *Naive*.

As we will show in the following, we develop an algorithm for computing the strict decisive skyline set with two salient features. First, our algorithm avoids evaluating all subspace skyline queries, and instead evaluates only $m + 1$ skyline queries. Second, assuming that data is indexed by a multidimensional index, we define an appropriate query that allows our algorithm to traverse the index at most once, retrieve a set of candidate points, and refine the result set in main-memory.

One important observation is that the strict decisive skyline points ($p \in \mathcal{DS}(P)$) are exactly those skyline points ($p \in \mathcal{S}(P)$) that for any $(m-1)$-dimensional subspace U they are either dominated ($p \notin \mathcal{S}_U(P)$) or share the same values in U with another data point p' ($p' =_U p$ and $p' \neq p$). We denote $p' =_U p$, if it holds that $\forall d_i \in U : p[i] = p'[i]$. However, in the simplest case one skyline query and m subspace skyline queries need to be processed and the index must be accessed multiple times. To avoid this processing overhead, we identify a super-set of this set that can be efficiently retrieved by traversing the index structure at most once.

Definition 7. *(Enriched skyline) A point $p \in P$ is said to partially dominate another point $q \in P$ on $\mathcal{D}$, if (1) on every dimension $d_i \in \mathcal{D}$, $p[i] \leq q[i]$; and (2) on at least two dimensions $d_j, d_k \in \mathcal{D}$, $p[j] < q[j]$ and $p[k] < q[k]$. The enriched skyline of P is the set of points $e\mathcal{S}(P) \subseteq P$ which are not partially dominated by any other point.*

The above definition assumes that the enriched skyline is defined on a data space that contains at least two dimensions, i.e., $|\mathcal{D}| \geq 2$. An interesting observation is that the enriched skyline uses a slightly modified definition of dominance, that can be supported by any skyline algorithm with marginal overhead, by simply changing the function used for point dominance. We can prove that the enriched skyline set is sufficient to compute the strict decisive skyline set $\mathcal{DS}(P)$.

3.1   Algorithmic Description

We design an efficient algorithm, called *Decisive Skyline Algorithm* and denoted as *DSA*, for computing the strict decisive skyline $\mathcal{DS}(P)$ of a set of points P. The innovative features of our algorithm include that (a) DSA operates only on a subset of the data set P, namely the enriched skyline set, that is both easy to compute and guaranteed to include all decisive skyline points, and (b) DSA computes the decisive skyline by efficient processing of the underlying subspace skyline queries without the need to access the disk repeatedly. DSA extends the well-known branch-and-bound (BBS) algorithm for skyline queries [14].

Algorithm 1. *Decisive Skyline Algorithm (DSA)*

---

input: The R-tree index $\mathcal{R}$ built on data set P
output: The decisive skyline set $\mathcal{DS}(P)$

1: $\mathcal{M} \leftarrow null$, $\mathcal{B} \leftarrow \emptyset$ //$\mathcal{M}$:main-memory R-tree, $\mathcal{B}$:buffer
2: $(\mathcal{S}(P), \mathcal{M}) \leftarrow BBS(\mathcal{R}, \mathcal{D})$
3: **for** $i = 1...m$ **do**
4:    $U \leftarrow \mathcal{D} - d_i$ //U:current subspace of $m-1$ dimensions
5:    $tmpDist \leftarrow -1$
6:    **while** has next point $BBS(\mathcal{M}, U)$ **and** $\mathcal{S}(P) \neq \emptyset$ **do**
7:       $q \leftarrow$ next point of $BBS(\mathcal{M}, U)$
8:       **if** $Dist_U(q) = tmpDist$ **then**
9:          $\mathcal{B} = \mathcal{B} \cup q$
10:       **else**
11:          **for all** $p \in \mathcal{B}$ **do**
12:             **if** $(p \in \mathcal{S}(P))$ **and** $(\nexists p' \in \mathcal{B} : \forall d_k \in U\ p[k] = p'[k]$ and $p[i] \neq p'[i])$ **then**
13:                $\mathcal{S}(P) \leftarrow \mathcal{S}(P) - p$
14:             **end if**
15:          **end for**
16:          $\mathcal{B} = \{q\}$
17:       **end if**
18:       $tmpDist \leftarrow Dist_U(q)$
19:    **end while**
20: **end for**
21: **return** $\mathcal{S}(P)$

The pseudocode describing DSA is shown in Algorithm 1. The algorithm takes as input a data set P indexed by an R-tree $\mathcal{R}$ and produces the decisive skyline set $\mathcal{DS}(P)$ as output. First, BBS is executed on the R-tree that indexes P, and it populates the main-memory R-tree $\mathcal{M}$ with the enriched skyline points (and only those). In addition, the skyline set $\mathcal{S}(P)$ is retrieved (line 2). Notice that the two parameters of the $BBS()$ call in the pseudocode correspond to the index used by BBS and the subspace which is processed respectively. Then, DSA executes m subspace skyline queries on the main-memory R-tree $\mathcal{M}$ iteratively (lines 3–20). For each $(m-1)$-dimensional subspace U, DSA exploits the progressive property of BBS and retrieves subspace skyline points sorted by their distance to the origin in subspace U (line 7) and places them in a buffer $\mathcal{B}$ (line 9). This guarantees that points with identical values in U will be processed in a batch. Processing of a batch of points includes examining each point p in $\mathcal{B}$ and checking whether it is a candidate point (lines 11–15). If so, then we test if there exists another point p' with identical values on the $m-1$ dimensions of U, but different on the last dimension. If such a point p' does not exist, the point p can safely be discarded from the candidate points (line 13). The same procedure is repeated until all subspace skyline points are processed or the candidate list $(\mathcal{S}(P))$ gets empty (line 6).

DSA exploits the main-memory R-tree internally used by BBS, in order to efficiently compute the decisive skyline set. Thus, DSA takes practically *for free* the index structure that is constructed by BBS during the skyline computation, thereby making the subsequent execution of subspace skyline queries extremely efficient. Moreover, as all decisive skyline points belong to the skyline set, we modify further BBS, so that only skyline points are reported as result $(\mathcal{S}(P))$, even though the main-memory R-tree indexes the enriched skyline points.

In practice, DSA processes m subspace skyline queries (of dimensionality $m-1$) using the main-memory R-tree, for excluding non-decisive skyline points from the already computed skyline set by BBS. Notice that the main-memory R-tree indexes only the enriched skyline points, therefore the execution of subspace skyline queries is more efficient compared to processing on the entire data set P on disk.

4   Experimental Evaluation

In this section, we provide an experimental study of the decisive skyline query. All algorithms are implemented in Java and the experiments run on a machine with 2x Intel Xeon X5650 Processors (2.66 GHz), 128 GB.

4.1   Qualitative Study

We perform skyline analysis on data extracted from DBLP, in order to discover researchers with significant number of publications on a combination of conferences. The data set contains data that reflect DBLP entries before 15/10/2008. We use the authors as points represented in a multidimensional space defined by the number of publications in selected conferences (dimensions). Major conferences from different research areas are selected as criteria that need to be balanced. We underline the strict decisive skyline points, while relaxed decisive skyline points are shown using bold. Each researcher is represented as a 3d point with values equal to the number of publications for each of the selected conferences, and higher values are preferable.

Table 1. $\widehat{\mathcal{DS}}(P)$: in bold, $\mathcal{DS}(P)$: underlined.

Id	Name	SIGMOD	PODS	CIKM	DecSub()
1	**Divyakant Agrawal**	**14**	**7**	**11**	{S, P, C}
2	Jeffrey F. Naughton	29	9	1	{S, P}
3	Amr El Abbadi	7	8	12	{P, C}
4	Jiawei Han	26	0	8	{S, C}
5	Dan Suciu	14	15	2	{P, C}
6	Serge Abiteboul	15	24	0	{S, P}
7	Michael J. Carey	36	3	0	{S}
8	Jeffrey D. Ullman	17	16	0	{S, P}
9	**Divesh Srivastava**	**28**	**10**	**3**	{S, C}, {P, C}
10	Yehoshua Sagiv	8	29	1	{P}
11	**Christos Faloutsos**	**19**	**4**	**8**	{S, P, C}
12	Raghu Ramakrishnan	28	14	1	{S, P}
13	Surajit Chaudhuri	33	8	0	{S, P}
14	Philip S. Yu	18	1	18	{C}
15	David J. DeWitt	33	1	1	{S, C}

Table 1 shows the skyline set for {SIGMOD, PODS, CIKM}. *Divyakant Agrawal* and *Christos Faloutsos* are the strict decisive skyline points and they balance nicely all criteria, compared to the remaining skyline points. By inspecting the result set, we observe that several researchers do not truly balance all given criteria (dimensions). Instead, they may balance subsets of the dimensions only, but not all of them. For instance, *Jeffrey D. Ullman* nicely balances SIGMOD and PODS, but not CIKM. The same holds for both *Serge Abiteboul* and *Raghu Ramakrishnan*, who are also experts in data management. On the other hand, *Yehoshua Sagiv* is included in the result due to the extremely high number of PODS publications. The decisive skyline query manages to exclude these points from the result set, thus returning only points that truly balance all criteria. It is likely that if a user were interested in only a subset of the criteria, she would have posed a 2d query with only the criteria of interest instead. On the other hand, *Divesh Srivastava* will be included in the relaxed decisive skyline set, because he is a subspace skyline in subspaces {PODS, CIKM} and {SIGMOD, CIKM} that cover the full space, thus he manages to balance all given criteria.

4.2 Performance of Decisive Skyline Algorithm

In the following, we study the cost of computing the strict decisive skyline query (using DSA), compared with the cost of computing the skyline query (using the BBS algorithm). Although in this latter experiment DSA and BBS compute two different result sets, the experiment aims to answer the following interesting question: *how much is the overhead of computing decisive skyline points, compared to the computation of the skyline set?*

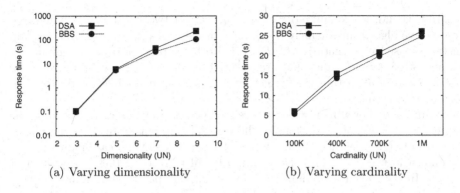

(a) Varying dimensionality   (b) Varying cardinality

Fig. 4. Comparative study: performance of DSA versus BBS.

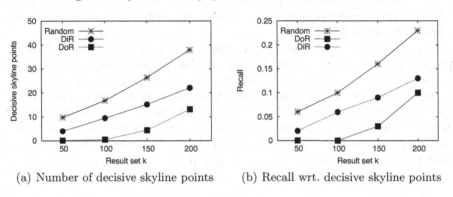

(a) Number of decisive skyline points   (b) Recall wrt. decisive skyline points

Fig. 5. Comparison with algorithms for representative skylines.

In Fig. 4(a), we study the effect of increased dimensionality, using a synthetic data set of 1M records following a uniform data distribution. We report the average results over 10 different instances of the data set. When increasing the dimensionality, the difference in response time between DSA and BBS is small and increases slowly.

In Fig. 4(b), we vary the cardinality from 100 K to 1 M, and we observe that the difference in time between DSA and BBS is small and constant, which is the expected result since the major impact in DSA is the number of subspace computations that is fixed in this setting. In summary, our finding is that DSA retrieves the strict decisive skyline set with a slightly increased cost compared to a state-of-the-art skyline algorithm (BBS [14]) that retrieves the traditional skyline set.

4.3   Comparison with Representative Skylines

Thereafter, we try to answer the following research question: *can the set of decisive skyline points be obtained by existing algorithms proposed for representative skyline computation?*

To this end, we compare our algorithm against two well-known skyline representative algorithms, namely dominance-based representative (*DoR*) [10] and distance-based representative (*DiR*) [17]. Both approaches take as a input the number k of skyline points that are selected as representatives. In addition, we use a *Random* algorithm that selects k representative skyline points from the skyline set at random.

Figure 5(a) shows the number of strict decisive skyline points retrieved by the representative skyline algorithms as the value of k increases. In this experiment, the size of strict decisive skyline set is 192 and the skyline cardinality is 952. As shown in the chart, both DoR and DiR fail to retrieve the decisive skyline points. In fact, even a random selection of skyline points (Random) retrieves more decisive skyline points than DoR and DiR. This demonstrates that the existing skyline representative algorithms do not take into account the semantics of the decisive skyline points and select completely different skyline points compared to our approach.

Figure 5(b) shows the recall that each representative skyline algorithm achieves, when using the strict decisive skyline points as correct result. As depicted in the chart, the recall of the representative skyline algorithms is very low, which demonstrates that these algorithms do not try (not even implicitly) to identify decisive skyline points. In fact, even a random selection of skyline points (Random) retrieves more decisive skyline points than DoR and DiR. This demonstrates that the existing skyline representative algorithms select completely different skyline points compared to our approach.

5   Conclusions

In this paper, we exploit the semantics of skyline points and propose a new query type, the *decisive skyline query*. Capitalizing on the concept of decisive subspaces, we define two variants of the decisive skyline set that are subsets of the skyline set. Points belong to the decisive skyline set due to their values in *all* user-specified criteria and provide interesting trade-offs. As a positive by-product and in contrast to skyline cardinality, the cardinality of the decisive skyline query is not significantly affected by dimensionality, thus leading to smaller result sets even for high-dimensional data. Our evaluation demonstrates the performance of our algorithm and that the decisive skyline query returns interesting points to the user.

In our future work, we intend to perform an in-depth study of the theoretical properties of decisive skyline queries and the cardinality of the decisive skyline set.

Acknowledgements. This work has been partly supported by the University of Piraeus Research Center.

References

1. Börzsönyi, S., Kossmann, D., Stocker, K.: The skyline operator. In: Proceedings of ICDE, pp. 421–430 (2001)

2. Chan, C.Y., Jagadish, H.V., Tan, K.L., Tung, A.K.H., Zhang, Z.: Finding k-dominant skylines in high dimensional space. In: Proceedings of SIGMOD, pp. 503–514 (2006)
3. Chan, C.Y., Jagadish, H.V., Tan, K.L., Tung, A.K.H., Zhang, Z.: On high dimensional skylines. In: Proceedings of EDBT, pp. 478–495 (2006)
4. Chaudhuri, S., Dalvi, N.N., Kaushik, R.: Robust cardinality and cost estimation for skyline operator. In: Proceedings of ICDE, p. 64 (2006)
5. Chomicki, J., Ciaccia, P., Meneghetti, N.: Skyline queries, front and back. SIGMOD Record 42(3), 6–18 (2013)
6. Ciaccia, P., Martinenghi, D.: Reconciling skyline and ranking queries. Proc. VLDB Endow. 10(11), 1454–1465 (2017)
7. Godfrey, P.: Skyline cardinality for relational processing. In: Seipel, D., Turull-Torres, J.M. (eds.) FoIKS 2004. LNCS, vol. 2942, pp. 78–97. Springer, Heidelberg (2004). https://doi.org/10.1007/978-3-540-24627-5_7
8. Hose, K., Vlachou, A.: A survey of skyline processing in highly distributed environments. VLDB J. 21(3), 359–384 (2012). https://doi.org/10.1007/s00778-011-0246-6
9. Lee, J., won You, G., won Hwang, S.: Personalized top-k skyline queries in high-dimensional space. Inf. Syst. 34(1), 45–61 (2009)
10. Lin, X., Yuan, Y., Zhang, Q., Zhang, Y.: Selecting stars: the k most representative skyline operator. In: Proceedings of ICDE (2007)
11. Lu, H., Jensen, C.S., Zhang, Z.: Flexible and efficient resolution of skyline query size constraints. IEEE TKDE 23, 991–1005 (2011)
12. Magnani, M., Assent, I., Mortensen, M.L.: Taking the big picture: representative skylines based on significance and diversity. VLDB J. 23(5), 795–815 (2014). https://doi.org/10.1007/s00778-014-0352-3
13. Mouratidis, K., Li, K., Tang, B.: Marrying top-k with skyline queries: relaxing the preference input while producing output of controllable size. In: Proceedings of SIGMOD, pp. 1317–1330 (2021)
14. Papadias, D., Tao, Y., Fu, G., Seeger, B.: Progressive skyline computation in database systems. ACM TODS 30(1), 41–82 (2005)
15. Pei, J., Jin, W., Ester, M., Tao, Y.: Catching the best views of skyline: a semantic approach based on decisive subspaces. In: Proceedings of VLDB, pp. 253–264 (2005)
16. Sarma, A.D., Lall, A., Nanongkai, D., Lipton, R.J., Xu, J.J.: Representative skylines using threshold-based preference distributions. In: Proceedings of ICDE, pp. 387–398 (2011)
17. Tao, Y., Ding, L., Lin, X., Pei, J.: Distance-based representative skyline. In: Proceedings of ICDE, pp. 892–903 (2009)
18. Vlachou, A., Vazirgiannis, M.: Ranking the sky: discovering the importance of skyline points through subspace dominance relationships. DKE 69(9), 943–964 (2010)
19. Xie, M., Wong, R.C., Lall, A.: An experimental survey of regret minimization query and variants: bridging the best worlds between top-k query and skyline query. VLDB J. 29(1), 147–175 (2020). https://doi.org/10.1007/s00778-019-00570-z
20. Yuan, Y., Lin, X., Liu, Q., Wang, W., Yu, J.X., Zhang, Q.: Efficient computation of the skyline cube. In: Proceedings of VLDB, pp. 241–252 (2005)
21. Zhang, Z., Yang, Y., Cai, R., Papadias, D., Tung, A.: Kernel-based skyline cardinality estimation. In: Proceedings of SIGMOD, pp. 509–522 (2009)

Safeness: Suffix Arrays Driven Materialized View Selection Framework for Large-Scale Workloads

Mohamed Kechar[1] and Ladjel Bellatreche[2(✉)]

[1] LITIO, Université Oran1 Ahmed Ben Bella, Ecole Superieure En Informatique,
Sidi Bel Abbes, Algeria
m.kechar@esi-sba.dz
[2] LIAS/ISAE-ENSMA, Poitiers, France
bellatreche@ensma.fr

Abstract. Materialized views (MVs) are an elegant redundant optimization technique for analytical workloads. Numerous frameworks and algorithms for selecting MVs have been propounded, and some are deployed in commercial DBMSs. The central role of MVs in optimizing workloads has recently prodded researchers to revisit the problems associated with them (selection, maintenance, adaptation) by proposing AI techniques to tackle those issues. When deeply analyzing both traditional and AI-driven solutions for the MV Selection Problem (VSP), we identify two main limitations: **(1)** the workloads used in experiments comprise dozens of queries, which is at odds with modern analytical projects involving large-scale workloads, and **(2)** Query Join Ordering (QJO) is not integrated into the VSP solving process. In this paper, we propose a framework named *Safeness* that tackles the VSP for large-scale workloads with the incorporation of QJO, thanks to suffix arrays initially introduced for string processing, enabling efficient algorithms for data compression, repeat finding, etc. Firstly, we show the flexibility of suffix arrays in coding analytical queries, capturing shareable subexpressions, and incorporating QJO. Secondly, an enumeration process of different MV candidates is specified. Thirdly, a cost-driven algorithm for selecting MVs under a storage budget is put forward. Finally, experiments are conducted to evaluate the effectiveness and scalability of our proposal.

Keywords: Materialized view · Suffix array · OLAP · Join order

1 Introduction

Materialized Views (MVs) are one of the most popular optimization techniques in the world of data stores. In addition to improving query performance, like classical optimization techniques (e.g., indexes, and data partitioning), MVs - if well selected - contribute to reducing *redundant computations* among queries [12]. This reduction is guaranteed by the smart exploitation of shareable sub-expressions that may exist among queries, especially in the context of modern analytical applications involving large-scale workloads [12]. The usage of MVs requires a substantial budget including storage cost and maintenance overhead [15]. Three main types of MVs exist (a) MVs with aggregates, (b) MVs containing only joins, and (c) nested MVs whose definition is based on other MVs and can reference base relation(s). The choice of the type of MV depends on the strategy of the materialization. The process of selecting a set of MVs is known as the MV Selection Problem (VSP). Its generic formalization has 5 inputs of which 4 are mandatory: (i) a datastore schema, (ii) the deployment infrastructure (centralized/distributed) hosting the datastore, (iii) a set of Non-Functional Requirements (NFR) (e.g., query performance, maintenance cost, energy consumption, number of final VMs), and (iv) a set of Constraints CsT (e.g., storage cost and pricing). The fifth one is optional and depends on the type of VSP (static/dynamic). It represents the workload. The VSP consists in selecting a set of MVs that optimizes NFR and satisfies C. VSP is *NP-Complete*, even in its simplest form [13]. Although, there exist numerous algorithms proposed by academia and industry dealing with the two types of VSP [15,18] in centralized and distributed deployment platforms [7,9,12], the majority, not to say all of them, of existing algorithms, uses the cost-driven vision widely adopted by both traditional and learned query optimizers [2,5,12,15,24,26]. Once MVs selected, initial workload queries are rewritten using views [11].

MVs and their surrounding problems (VSP, view maintenance, query rewriting) have proved to be a fertile field for research for many years. An in-depth analysis of the literature helps us identify three main periods: (a) *the "hot topic" period* characterized by researchers' race to tackle MV problems after the advent of data warehouses (DW). The presence of a session well displayed on MVs in the program of the DaWaK conference from 1999 until 2006 is a marker of this period. (b) The *"maturity" period*, in which we have seen major advances in developing new classes of algorithms [15], studying the interaction between MVs and other optimization techniques such as indexes and data partitioning [2,26], and the incorporation of MV selection algorithms in advisors of major DBMS [5,26]. *The proposed VSP algorithms in these two periods use workloads with dozens of queries* [12]. (c) The *"scalability" period*, characterized by the arrival of Big Data and the multiplicity of users consuming it. These considerations question the limitations of existing approaches to dealing with large-scale workloads.

To the best of our knowledge, [6] is the first academic work that discussed these limitations and recommended the usage of *hypergraph structures* to capture shareable subexpressions. They evaluated their proposal by considering 10,000 queries. From an industry perspective, the recent Microsoft paper [12] revives the VSP according to four aspects: (i) the availability of large-scale workloads including tens of thousands of jobs in modern shared analytics clusters (ii) the presence of *significant redundant computations* in jobs that can be reduced using VMs. This phenomenon has also been also observed in real workloads of Alibaba

Cloud [24]. **(iii)** The materialization of shareable subexpressions contributes to saving up to 40% machine-hours in Microsoft clusters [12]. This represents a spectacular saving in computation time [12] and as a consequence energy consumption [18]. **(iv)** Selecting MVs for large-scale workloads passes through a combination of two problems: Subexpression Selection (*PSS*) and *VSP* usually studied separately [25]. *PSS* aims at producing a pool of parts of logical queries. Its integration into VSP is highly recommended in analytical applications.

The two reference studies from both academia and industry [6,12] have had the merit to prod research to revisit VSP. By analyzing them, we evidence two important missing issues: (1) the usage of graph-driven modeling (AND-OR DAG [9], Multiple View Processing Plan (MVPP) [22], and Hypergraphs [16]) to detect shareable subexpressions. These structures were widely employed in periods 1 and 2. None one can dispute the fact that the construction and the exploitation of graphs are time-consuming and require a high memory budget. Distributed infrastructures have been used to overcome these drawbacks [12]. (2) The variation of Query Join Order (QJO) is not explicitly integrated into the process of solving VSP, despite the fact that it shows its impact on optimizing queries [21] and selecting shareable query subexpressions [23].

In this paper, we propose *Safeness* – a new MV selection framework for large-scale analytical workloads that involve joins across star and snowflake schemas, selections, sorting, grouping and aggregations, and nested sub-queries [8]. Contrary to existing frameworks, *Safeness* integrates QJO and uses *generalized suffix arrays* (*GESA*) known by their linear-time construction [14], and their capacity in enabling efficient algorithms for detecting repeat finding in large-scale texts, and plagiarism [4]. The usage of *GESA* in our context is not straightforward, since it requires a real adaptation in terms of *coding analytical queries, capturing shareable subexpressions, and using them* in our MV selection algorithm.

Our paper is organized as follows: Sect. 2 describes our related work. Section 3 shows how *GSA* is applied to analytical queries. Section 4 presents our suffix array-driven framework *Safeness* for selecting MVs for large-scale workloads. Section 5 validates our proposals. Section 6 concludes the paper.

2   Related Work

The process of selecting MVs borrows two major ideas from *PSS* [12] and multi-query optimization problem (*MQO*) [20,21]: (1) the usage of graph-based modeling and (2) exploration strategies. In addition, *VSP*, *PSS*, and *MQO* are quite similar and a few details distinguish them [12,23]. The *VSP* is much more general than *PSS* since the former can consider computations that do not appear in the workload queries. This increases the research space and complicates query containment and materialized view rewrites [3,12]. The main difference between *MQO* and *VSP* is that MVs are typically only transiently materialized for the execution of a given query set [12]. Due to the availability of surveys and encyclopedia papers related to these problems [15,19], we will not present their state-of-the-art, but we *deeply analyze them* to find missing issues that contribute to revisiting them in the context of large-scale analytical workloads.

VSP has been widely studied in the literature by academia and industry [2, 6, 9, 18, 26]. Each existing algorithm is associated to an instantiation of our generic formalisation (cf. Introduction). For instance, in [9], the authors consider query performance as a NFR and view maintenance overhead as a constraint. In [18], two NFR are considered to represent the performance and energy efficiency of queries. The work conducted by Oracle [3] considers three constraints: (a) the number of selected MVs must be small and with a reasonable size, (b) they must contain large pre-computations of joins and grouping, and (c) can rewrite a substantial number of current and future workload queries. It should be noticed that the resolution of the VSP depends on the result of PSS. This dependency is not always considered [12]. [22] is one of the pioneering works that identified the strong connection between VSP and PSS in the context of *small OLAP workloads*. The authors proposed an approach that consists in analyzing the queries so as to derive common intermediate results which can be shared among the queries. An 0–1 integer programming algorithm is used to generate optimal MVPPs. It uses single rule-based query optimizations combined with query tree merging techniques which aims to incorporate the individual optimal query plans as much as possible in the MVPP. In [12], the authors focused on PSS for large workloads and proposed an ILP-based formulation of VSP. The BIGSUBS algorithm is given to select subexpressions to materialize. Its main idea is to split the initial problem into two subproblems, where each one is solved separately in an iterative manner. Several pruning strategies have been proposed. A distributed implementation of BIGSUBS using SCOPE has been proposed [12]. VSP, PSS, and MQO are more sensitive to QJO that may impact dramatically the query sharing [21]. Their respective solutions assume that QJO is known (e.g., usually they use the one delivered by the target datastore [24]).

Another finding of our analysis of existing works is that the management of joins/projections/selections is not always. Two transformations are distinguished: (i) joins have priority over selections and projections, and (ii) all operations have the same priority. In the first transformation, selections/projections are removed and put back after merging queries. Different selections defined on a table are performed in a disjunctive form [22]. This transformation may increase the number of shareable join subexpressions with large sizes. In the second management, queries are used without any transformation. This may decrease the number of shareable join subexpressions [12, 16].

3 Suffix Arrays for Large-Scale Workload Coding

Due to the novelty of using suffix arrays (SA) in the process of selecting MVs, a presentation of its fundamental concepts is necessary. After that, we show how SA is reproduced in large-scale workloads management.

3.1 Fundamental Concepts of Suffix Arrays

Definition 1. *Let w be a string with length n defined over an alphabet $\Sigma \cup \{\#\}$, where $\#$ represents a special symbol (called sentinel) and not belonging to Σ.*

All symbols in Σ are larger than #. The j-th $(1 \leq j \leq n)$ suffix of the w is a substring $[j \ldots n]$. The suffix associated to the whole string is called full suffix. The SA of w (denoted by SA_w) is an array of integers specifying the starting positions of suffixes of w in a lexicographical order. Only the indices of suffixes are stored in the SA_w instead of whole suffixes satisfying the following property: $\forall i \in 1 \ldots n : w[SA[i-1] \ldots n] < w[SA[i] \ldots n]$

The SA of "banana#" and "banned#" are given respectively in Fig. 1 (a,b). The SA of a string w is enhanced with the Longest-Common-Prefix (LCP) array which forms Enhanced Suffix Array (ESA) [1]. An LCP contains the lengths of the longest common prefixes of adjacent suffixes in SA_w. More formally, let $lcp(u, v)$ be the length of the longest common prefix of strings u and v, the LCP array of u is an array of integers such that $LCP[k] = lcp(u[SA[k-1] \ldots n], u[SA[k] \ldots n])$ $(\forall k \in 2 \ldots n)$ and LCP[1] = 0. The ESA of "banana#" and "banned#" is given in Fig. 1 (a, b). Till now, we only considered suffix array of a single string. In real life applications suffix arrays are generalized to code a *large collection of strings* [1]. To do so, let us consider a collection of N strings $CS = \{w_1, w_2, ..., w_N\}$ defined over an alphabet $\Sigma \cup \{\#\}$. Each string w_i $(1 \leq i \leq N)$ is associated to a length n_i. It is also *indexed* by an integer representing its position in CS (called *string position in CS*). Let T be the concatenation of all N strings: $w_1\#w_2\#...\#w_N\#$, and separated by #.

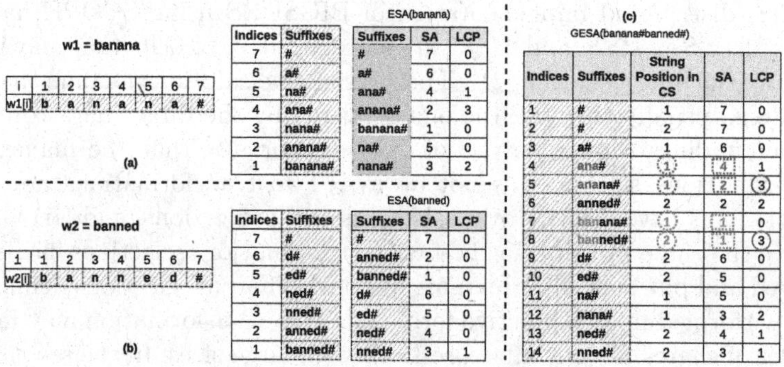

Fig. 1. An example of GESA of *banana* and *banned*

Definition 2. *The Generalized Enhanced Suffix Array (GESA) of T is an array specifying the lexicographic ordering of all suffixes of T. It contains pairs of integers (i, j) that specifies the lexicographic order of all suffixes $w_i[j, n_i]$ of strings in CS $(1 \leq i \leq N$ and $j = SA[h]$ for $h = 1 \ldots |T|)$.*

The $GESA$ of T composed by "banana#" and "banned#" is represented by a rich data structure composed by 5 entries: **(a)** indices representing all suffixes of T, **(b)** suffixes of T, **(c)** a string position in CS, **(d)** SA, and **(e)** LCP (Fig. 1 (c)). The LCP entry is crucial for detecting redundant substrings over intra-string or inter-strings. In intra-string case, the most popular studied problem is the detection of longest repeated substring of a string [10]. The highest value in LCP (=3) (represented by a blue circle) in Fig. 1 gives the length of the longest repeated substring $ana\#$ (of $banana\#$). When several strings are considered, detecting longest common substrings is largely studied [10], which corresponds exactly to our studied problem.

Definition 3. *A footprint substring represents the longest common substring among several strings.*

"*ban*" gives the footprint of $banana\#$ and $banned\#$.

3.2   Application of Suffix Arrays to OLAP Queries

To code an OLAP query by a SA, an important hypothesis is given: selections and projections are removed and put back in a push-down fashion once the MVs are identified. To illustrate this coding let us consider the following example.

Example 1. Before detailing our example, we first, we overview Yang et al. approach [22] (*named Yang_A*) considered as a reference for solving the combined *PSS* and *VSP*. Let us consider three-star schema queries (Q_1, Q_2, Q_3) defined on Star Schema Benchmark (SSB) composed of one fact table: *Lineorder* (LO), and 4 dimension tables: *Date* (DA), *Customer* (Cu), *Part* (PA), and *Supplier* (SU). Due to the lack of space, we consider only their algebraic expressions. We assume that QJO is a priori known.

$$Q_1 : \Pi_{(p_1,p_2,p_3)}\left[\sigma_{(cl_2 \wedge cl_1 \wedge cl_3)}(LO \bowtie DA \bowtie SU \bowtie CU)\right];$$
$$Q_2 : \Pi_{(p_4,p_5,p_6,p_7)}\left[\sigma_{(cl_5 \wedge cl_6 \wedge cl_4 \wedge cl_3)}(LO \bowtie DA \bowtie SU \bowtie PA \bowtie CU)\right]; \text{ and}$$
$$Q_3 : \Pi_{(p_8,p_7)}\left[\sigma_{(cl_7 \wedge cl_3)}(LO \bowtie DA \bowtie CU)\right].$$

Yang_A starts by constructing left deep trees of the queries by pushing up selection operations (Fig. 2 (a). Secondly, these trees are merged by exploiting shareable nodes (joins). This merging produces a unified graph called MVPP (Fig. 2 (b)). Selections are pushed as far down as possible in the MVPP. Two types of intermediate nodes are distinguished: **(i)** a node with fanout equal to 1 that represents a subquery of a query (the case of J_2 and J_3) and **(ii)** a node with fanout greater than 1 that represents at the same time a subquery and shareable subexpression (ex. nodes J_0 and J_1). If a node is shared by many queries of a workload, its materialization may contribute to avoiding redundant computations (J_0 and J_1).

To use SA to code our queries, we first generate our Alphabet $\Sigma = \{0, 1, 2, 3, 4, \#\}$ that includes integers assigned to each table: $LO \leftarrow 0, DA \leftarrow 1, SU \leftarrow 2, PA \leftarrow 3, CU \leftarrow 4$). Then, each query is coded by a string. For instance, Q_1, after ignoring selections/projections, it becomes $LO \bowtie DA \bowtie SU \bowtie CU$, is coded as "0124" (Fig. 2 c). Each suffix starting from position 1 of any query string with a length greater than 2 represents a subexpression of that query (Fig. 2 d). The concatenation of all strings related to our queries produces the string $T =$ "0124#01234#014#". Its GESA is described in Fig. 2

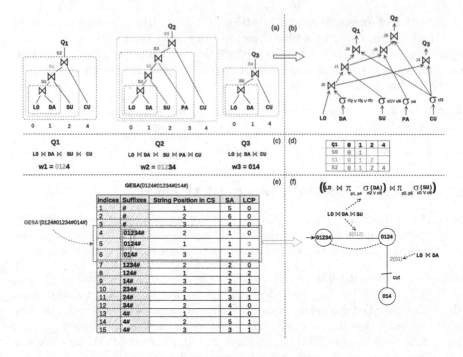

Fig. 2. Graph vs. Suffix arrays

(e). It can be used to identify shareable subexpressions and enable defining MV selection algorithms based on the target NFR based on hints: (1) if query performance overrides the others NFR, the materialization of the longest common substrings among queries ("012" in our case) could be the best choice. The materialization of "012" optimizes only Q_1 and Q_2 (by saving 2 joins), and saves also 2 computations. (2) In the case that the reduction of the number of redundant computations among queries is privileged, the materialization of any common substring ("01" in our case) represents the best choice. The materialization of "01" contributes to optimizing all queries (1 join is saved), and saving 3 computations. Any common substring is called *footprint join subexpression*. Once footprint join expressions are identified, selections and projections are put back. The obtained expressions are called *covering subexpressions*. The longest common substring after varying the QJO of Q_1 ("0142") and Q_2 ("01423") is "014". The materialization of "014" optimizes all queries and saves 3 computations. This shows the central role of QJO in PSS and VSP solving.

4   Safeness Framework for MV Selection

In this section, we present our framework *Safeness* to select MVs for a large scale workload defined on a DW composed of a fact table F and L dimension tables $\{D_1, \ldots, D_L\}$. *Safeness* is composed of 7 components:

1. **Query parsing:** Each SQL query of the workload is parsed in order to extract its corresponding tables, selections, projections, and joins. Let $Sel^{Q_i}_{D_j}$ be the set of selections of the query Q_i defined on the dimension table D_j.

2. **QJO module:** Since one of our objectives is to measure the impact of QJO on the VSP, we define three QJO strategies: (i) QJO delivered by PostgreSQL[1], Min2Max, and Fanout QJO. The hint behind the two last QJO is to minimize the size of the intermediate result of the join. In Min2Max QJO, joins are performed according to their sizes in ascending order, by ignoring the selections that may involve dimension tables. Fan-out integrates these selections and evaluates their effect on the fact table F in terms of reducing its size. More precisely, for a query q having a selection predicate p defined on a dimension table D, the $fan\text{-}out(p)$ is computed as follows: $Fan\text{-}out(p) = \frac{\|F \ltimes \sigma_p(D)\|}{\|F\|}$, where $\| \ \|$ and $\ltimes$ represent respectively the number of instances of a table, and semi-join operation.

3. **Construction of GESA:** This is done for each QJO strategy. The construction follows the same principle developed in Example 1 and thanks to the $eGSA$ algorithm [17] chosen for its efficiency guaranteed by the usage of external memory when dealing with large collections.

4. **Footprint Join Subexpressions Generation:** This step consists in capturing common substrings among query strings while integrating the semantics of analytical queries stipulating that the fact table must occupy either the first or second positions in any query string. Let $GESA'$ be a fragment of the $GESA$ structure. It contains only the set of full suffixes. The problem of finding footprint subexpressions is equivalent to the classification of full suffixes in classes, where each class contains full suffixes having the same value of LCP. Figure 2 (f) shows the result of the classification, where two classes are well identified, and separated by a cut. For each class, its footprint join expression is generated. The substring "012" in Fig. 2 (f) is translated to a join sequence $LO \bowtie DA \bowtie SU$.

5. **Covering Subexpressions Generation:** Let $\mathcal{FP} = \{fp_1, fp_2, \ldots, fp_b\}$ be a set of all footprints obtained in step 4. Let Q^{fp_j} is a set of queries sharing fp_j. Each footprint is transformed to a covering expression by pushing down selections defined on each table in disjunctive way. The covering join subexpression $\big((LO \bowtie \Pi_{(p_1,p_4)}\sigma_{(cl_2 V cl_5)}(DA)) \bowtie \Pi_{p_2,p_5}\sigma_{cl_1 V cl_6}(SU)\big)$ shown in Fig. 2 (f) corresponds to the footprint subexpression $LO \bowtie DA \bowtie SU$ after pushing down selections and projections.

6. **Benefit of covering subexpressions over the workloads:** Since, we are following a cost-based approach for selecting our MVs, we have to quantify the benefit of each covering subexpression cov_j over each query of $q_i^{fp_j} \in Q^{fp_j}$ (denoted by $B^{cov_j}(q_i^{fp_j})$). It is computed as follows: Let s_j be the subquery of $q_i^{fp_j}$ that matches fp_j. $B^{cov_j}(q_i^{fp_j}) = PC(s_j) - AC(cov_j)$, where $PC(s_j)$ and $AC(cov_j)$ represent respectively the processing cost of s_j and the cost

---

[1] https://www.postgresql.org/docs/8.1/geqo-pg-intro.html.

of accessing cov_j (if materialized). The total benefit of cov_j over Q^{fp_j} is:
$B^{cov_j}(Q^{fp_j}) = \sum_{q_i^{fp_j} \in Q^{fp_j}} B^{cov_j}(q_i^{fp_j})$.

7. **MV Selection:** The previous steps gives a set of covering subexpressions $\{cov_1, cov_2, \ldots, cov_b\}$, where each cov_i has a total benefit B_i and a size z_i. All are MV candidates. By integrating the storage constraint C, our VSP can be modeled as a knapsack problem and reduced to Integer Linear Programming (ILP) as follows: $maximize(\sum_{i=1}^{b} B_i x_i)$; subject to $\sum_{i=1}^{b} z_i x_i \leq C$, $x_i \in \{0,1\}$, $i = 1, \ldots, b$

5 Experimental Study

In this section, we conduct intensive experiments by considering several large-scale workloads to evaluate the effectiveness, the scalability, and the capacity of detecting redundant computations of our proposal. Yang_A [22] a graph-based algorithm chosen as a baseline solution. For these experiments, we use a desktop machine with 8 GB of RAM and 4 cores of 2.4 GHz processors. We build a synthetic workload generator based on the query templates of the SSB. Four workloads with different sizes (10 000, 20 000, 30 000, and 40 000) are generated and executed over a DW with 102 000 000 facts. The number of joins in each workload varies from 1 to 4. The evaluation and access costs of each shareable and covering subexpressions are measured by the PostgreSQL query optimizer. *Safeness* is implemented using Python. The Python-embedded modeling language for convex optimization problems[2] is used to solve our ILP problem.

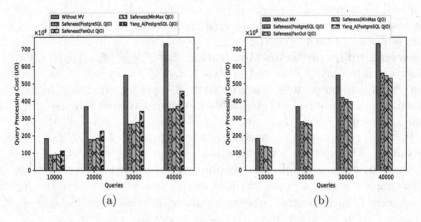

Fig. 3. (a) QP (no storage limit), (b) QP (100% storage)

---

[2] https://www.cvxpy.org/.

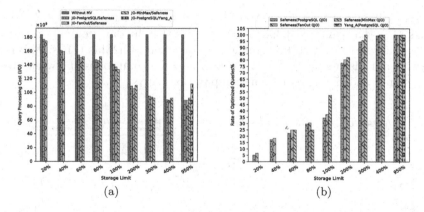

Fig. 4. (a) QP by varying the storage, (b) Rate of optimized queries

Effectiveness. To evaluate the effectiveness of *Safeness*, we measure the query performance (QP) of each workload with and without MVs whilst varying join ordering strategies. In the first experiment, the storage constraint is not imposed. Therefore, all covering subexpressions generated by our algorithm are materialized. The obtained results are depicted in Fig. 3 (a). They show that whatever the size of the workloads, the MVs obtained by PostgreSQL and Fanout QJO give a better improvement than those obtained by Min2Max QJO. Another important result concerns our superiority of *Safeness* over Yang_A in terms of scalability. For instance, for 40 000 queries, *Safeness* achieves improvement rates of 51.51%, 51.48%, and 49.8% corresponding respectively to the PostgreSQL, Fanout, and Min2Max QJO, whereas Yang_A realizes 37.7% with the PostgreSQL QJO. In our second experiment, we consider the storage constraint. Figure 3 (b) illustrates the QP costs achieved by *Safeness* under a *storage reference constraint* equal to the total size of the underlying *DW*. Whatever the used QJO strategy and the size of the workloads, *Safeness* is able to select beneficial MVs respecting the storage budget, whereas Yang_A completely fails in selecting MVs. Yang_A starts selecting MVs when the storage limit reaches 9 times the size of the *DW* (Fig. 4 (a)). The trade-off between QP improvement and the storage constraint satisfaction is mainly due to the fact that *Safeness* captures the longest footprint subexpressions common to each class of queries. Covering subexpressions of that class build upon these longest footprint subexpressions, if materialized, have the smallest sizes. This is because their joins are performed on filtered tables by pushing down only selections used by the class of queries. The performance of these queries and redundant computation savings are inevitably guaranteed. In Yang_A, all intermediate nodes in MVPP are MV candidates whatever their lengths. Moreover, these candidates are defined on tables, where each one is filtered by pushing down its selections used by all queries, which increases the size of MV candidates. This is the reason why Yang_A fails in selecting MVs for large workloads under a reasonable storage budget. Figure 4 (b) depicts the saved computations achieved by *Safeness* and Yang_A for a workload of 10 000 queries and storage limit equal to the size of the *DW*. For instance, *Safeness* selects MVs that can save 3 495, 3 730, and 5 241 redundant computations corresponding respectively to PostgreSQL, Fanout, and Min2Max QJOs. This is in line with our discussion in Sect. 3.2 concerning the compromise between query performance and saving redundant computation.

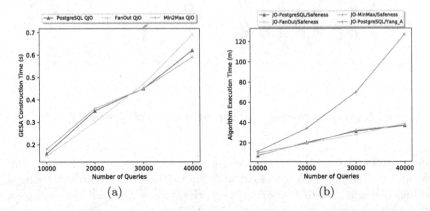

Fig. 5. (a) GESA construction time, (b) Run-time of the VS Algorithm

Scalability. To assess the scalability of *Safeness*, we report the runtime of the construction of the *GESA* (Fig. 5 (a)) and the runtime of both *Safeness* and Yang_A by varying our workloads (Fig. 5 (b)). Figure 5 (a) illustrates the time-linearity of the construction algorithm of the *GESA*. More interesting, the construction time is very small and is about 0.7 s for a workload of 40 000 queries. Figure 5 (b) depicts the executions time of *Safeness* with three QJO strategies, as well the execution time of Yang_A by considering only one MVPP produced by merging query trees using the PostgreSQL QJO. As we can observe, *Safeness* scales up whenever the workload size grows and maintains a runtime linear to the size of the workload. However, the execution time of the Yang_A is proportional to the number of queries, which penalizes its usage for large-scale workloads. For instance, for the workload of 40 000 queries (with the PostgreSQL QJO), our algorithm takes 37 min (on a single and classical desktop machine) to deliver the best MVs that contribute to improving QP up to 51% (Fig. 3 (a)), while Yang_A requires more than 2 h to perform its MV selection with QP improvement rate equal 37.7%.

Impact of QJO on MV Selection Algorithms. Our experiments show clearly the real impact of QJO on the whole process of selecting MVs (Fig. 4 (a)). For instance, for a workload of 10 000 queries, *Safeness* selects MVs that reduce QP up to 3.75% and 4.93% for a 20% of storage reference constraint, and 12.47% and 13.22% for 40% of storage reference limit, when using respectively the PostgreSQL and FanOut QJO. However, under these storage constraints, the Min2Max is an inappropriate QJO strategy since *Safeness* fails in selecting any MV. We observe also that when the storage limit varies, the quality of the selected MVs varies, in turn, depending on the used join ordering. This result is spectacular and reinforces our initial claim about the strong dependency between QJO, VSP, MQO, and PSS.

6 Conclusion

In this paper, we revisited VSP in order to optimize the performance of modern analytical applications known for their large-scale workloads and to reduce redundant computations. We took advantage of this revisit to integrate QJO in the whole process of selecting MVs. This revisit allows both delivering a historical view of the evolution of MVs and their surrounding problems, analyzing the state-of-the-art of VSP, MQO, PSS, and managing joins, selections, and projections when dealing with VSP. We believe that these findings will motivate researchers to continuously revisit the VSP in the era of large-scale workloads. To cope with the volume of queries, we substituted graph-driven modeling of queries with the *GESA* data structure known for its efficiency in text processing. Contrary to existing studies that consider the entire workload to materialize views, our framework *Safeness* partitions it into several classes, where each one contains queries sharing the longest subexpressions. The materialization of these subexpressions will naturally optimize large-scale workloads and save redundant computations. Our preliminary results are encouraging and are justified by the consideration of dependencies among PSS, VSP, and QJO.

The nice ideas presented in this paper about the consideration of QJO in VSP, the usage of GESA in coding analytical queries, capturing shareable subexpressions and saving redundant computations among queries, open several directions: **(a)** the implementation of recent AI algorithms for QJO and evaluation of their impact on the selected MVs. **(b)** Providing an adaptive version of *Safeness*, by considering the significant changes in the target workload. **(c)** The development of GESA driven algorithms for selecting indexes for large-scale workloads.

References

1. Abouelhoda, M.I., Kurtz, S., Ohlebusch, E.: Replacing suffix trees with enhanced suffix arrays. J. Discrete Algorithms **2**(1), 53–86 (2004)
2. Agrawal, S., Chaudhuri, S., Narasayya, V.R.: Materialized view and index selection tool for microsoft SQL server 2000. In: ACM SIGMOD, p. 608 (2001)
3. Ahmed, R., Bello, R.G., Witkowski, A., Kumar, P.: Automated generation of materialized views in oracle. VLDB Endow. **13**(12), 3046–3058 (2020)
4. Barenbaum, P., Becher, V., Deymonnaz, A., Halsband, M., Heiber, P.A.: Efficient repeat finding in sets of strings via suffix arrays. Discret. Math. Theor. Comput. Sci. **15**(2), 59–70 (2013)
5. Belknap, P., Dageville, B., Dias, K., Yagoub, K.: Self-tuning for SQL performance in oracle database 11g. In: ICDE, pp. 1694–1700 (2009)

6. Boukorca, A., Bellatreche, L., Senouci, S.-A.B., Faget, Z.: SONIC: scalable multi-query optimization through integrated circuits. In: Decker, H., Lhotská, L., Link, S., Basl, J., Tjoa, A.M. (eds.) DEXA 2013. LNCS, vol. 8055, pp. 278–292. Springer, Heidelberg (2013). https://doi.org/10.1007/978-3-642-40285-2_24

7. Chaves, L.W.F., Buchmann, E., Hueske, F., Böhm, K.: Towards materialized view selection for distributed databases. In: EDBT, pp. 1088–1099 (2009)

8. Chen, J., Jindel, S., Walzer, R., Sen, R., Jimsheleishvilli, N., Andrews, M.: The MemSQL query optimizer: a modern optimizer for real-time analytics in a distributed database. VLDB Endow. 9(13), 1401–1412 (2016)

9. Gupta, H., Mumick, I.S.: Selection of views to materialize in a data warehouse. IEEE Trans. Knowl. Data Eng. 17(1), 24–43 (2005)

10. Gusfield, D.: Algorithms on Strings, Trees, and Sequences - Computer Science and Computational Biology. Cambridge University Press, Cambridge (1997)

11. Halevy, A.Y.: Answering queries using views: a survey. VLDB J. 10(4), 270–294 (2001). https://doi.org/10.1007/s007780100054

12. Jindal, A., Karanasos, K., Rao, S., Patel, H.: Selecting subexpressions to materialize at datacenter scale. VLDB Endow. 11(7), 800–812 (2018)

13. Karloff, H.J., Mihail, M.: On the complexity of the view-selection problem. In: PODS, pp. 167–173 (1999)

14. Louza, F.A., Telles, G.P., Hoffmann, S., de Aguiar Ciferri, C.D.: Generalized enhanced suffix array construction in external memory. Algorithms Mol. Biol. AMB 12, 1–16 (2017)

15. Mami, I., Bellahsene, Z.: A survey of view selection methods. SIGMOD Rec. 41(1), 20–29 (2012)

16. Mouna, M.C., Bellatreche, L., Boustia, N.: ProRes: proactive re-selection of materialized views. Comput. Sci. Inf. Syst. (2022)

17. Puglisi, S.J., Smyth, W.F., Turpin, A.H.: A taxonomy of suffix array construction algorithms. ACM Comput. Surv. 39(2), 4-es (2007)

18. Roukh, A., Bellatreche, L., Bouarar, S., Boukorca, A.: Eco-physic: Eco-physical design initiative for very large databases. Inf. Syst. 68, 44–63 (2017)

19. Roy, P., Sudarshan, S.: Multi-query optimization. In: Liu L., Özsu, M.T. (eds.) Encyclopedia of Database Systems, 2nd Edition. Springer, New York (2018). https://doi.org/10.1007/978-1-4614-8265-9_239

20. Sellis, T.K.: Multiple-query optimization. ACM Trans. Database Syst. 13(1), 23–52 (1988)

21. Sioulas, P., Ailamaki, A.: Scalable multi-query execution using reinforcement learning. In: ACM SIGMOD, pp. 1651–1663 (2021)

22. Yang, J., Karlapalem, K., Li, Q.: Algorithms for materialized view design in data warehousing environment. In: VLDB, pp. 136–145 (1997)

23. Yu, X., Li, G., Chai, C., Tang, N.: Reinforcement learning with tree-LSTM for join order selection. In: ICDE, pp. 1297–1308 (2020)

24. Yuan, H., Li, G., Feng, L., Sun, J., Han, Y.: Automatic view generation with deep learning and reinforcement learning. In: ICDE, pp. 1501–1512 (2020)

25. Zhou, J., Larson, P., Freytag, J.C., Lehner, W.: Efficient exploitation of similar subexpressions for query processing. In: ACM SIGMOD, pp. 533–544 (2007)

26. Zilio, D.C., et al.: DB2 design advisor: integrated automatic physical database design. In: VLDB, pp. 1087–1097 (2004)

A Process Warehouse for Process Variants Analysis

Lisana Berberi[✉][ID]

Karlsruhe Institute of Technology, Hermann-von-Helmholtz-Platz 1,
76344 Karlsruhe, Germany
lisana.berberi@kit.edu
https://www.scc.kit.edu

Abstract. Process model variants are collections of similar process models evolved over time because of the adjustments that were made to a particular process in a given domain, e.g., order-to-cash or procure-to-pay process in reseller or procurement domain. These adjustments produce some variations between these process models that mainly should be identical but may differ slightly. Existing approaches related to data warehouse solutions suffer from adequately abstracting and consolidating all variants into one generic process model, to provide the possibility to distinguish and compare among different parts of different variants. This shortcoming affects decision making of business analysts for a specific process context. This paper addresses the above shortcoming by proposing a framework to analyse process variants.

The framework consists of two original contributions: (i) a novel meta-model of processes as a generic data model to capture and consolidate process variants into a reference process model; (ii) a process warehouse model to perform typical online analytical processing operations on different variation parts thus providing support to decision-making through KPIs; The framework concepts were defined and validated using a real-life case study.

Keywords: Process variant · Process warehouse · Business process analysis

1 Introduction

Process model variants, as collections of similar process models, may evolve over time because of the adjustments made to the same business process in a given domain, e.g., order-to-cash or procure-to-pay process in reseller or procurement domain. These adjustments produce some variations between these process models. Surely, between business processes across department of the same organization, or across companies in a given industry many common activities are frequently found. For example, typical process *procure-to-pay* often consists of a business process that starts from the moment a procure invoice is received from

R. Wrembel et al. (Eds.): DaWaK 2022, LNCS 13428, pp. 87–93, 2022.
https://doi.org/10.1007/978-3-031-12670-3_8

a vendor after a customer places an order and fulfilled if the vendor has received the corresponding payment. All these *procure-to-pay* processes include activities related to receiving, invoicing and payment. They may look the same but they slightly differ from each other. Especially, of great importance is having an information on management of the work progress between different parts of different variants and then select the most efficient one. Dedicated technologies lack on effectively manage the information on processes encoded in process models and process execution records [18]. For more than a decade process-oriented data warehouse have been introduced as a solution on analysing effectively process activities of an organization. Process Warehouses [2,9,17] are an appropriate means for analysing the performance of business process execution using well established data warehouse technology and on-line analytical processing (OLAP) tools. A way to manage these variants is expressing all the variants in a single process definition with the excessive use of XOR-Splits. The resulting processes are large, difficult to understand and overloaded, and new process definitions still comprise of all the past processes definitions they should replace.

To address these shortcomings we propose a process warehouse model which allows to express a generalization hierarchy of processes to adequately capture process variants. This generalization hierarchy can be generated from a meta-model of business process models which introduces the notion of generic activities which generalize a set of activities (e.g., pay by credit card, by check, or by third-party (PayPal) could all be generalized to an activity payment). Based on these given hierarchies of activities we can define generalization hierarchy of processes for the "process" dimension of a process warehouse. This hierarchy can then be used to roll-up or drill down when analysing the logs of the executions of the various process variants and it makes it much easier to compare key-performance indicators between different variants at different levels of genericity. In this context, the main research question this paper addresses is:

RQ: How can a family of process variants be effectively and efficiently analysed using a process warehouse approach?

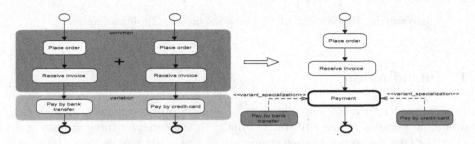

Fig. 1. An example of reference (global) process model abstracting multiple process variants

This research question specifies the interoperability between business process modelling, enactment and data warehouse research areas with the aim of analysing different variants in a multidimensional perspective. To identify how effective (e.g. measure customer satisfaction for a product or process) and efficient (e.g. measure time, cost and resource utilization) a business process is, a process performance analysis is crucial. Moreover it helps in estimating process improvement efforts. To understand how a reference or global process model is constructed, let us consider a concrete example that refers to customer invoice payments after ordering his/her goods or services. Figure 1 shows two variants of the order-to-pay process represented Business Process Modelling Notation (BPMN) [7]. These variants reflect two possibilities to pay: the first pay by bank transfer (filling a bank statement), the other, pay by credit-card (check customer balance). We show how a reference can be constructed by identifying the commonalities and variability among them. The choice between pay by credit-card or pay by bank transfer represents a variability in this process: depends on different drivers such as type of invoice, type of goods etc. The two variant activities are integrated to a new generic (abstract) activity named *Payment* as shown on the right-hand side of the figure. We use a stereotype named ≪*variant_specialization*≫ assigned to the generic connector between the generic activity and the specialized activities. We present our method to deal with process variants, specifically we design a meta-model to adequately capture process variants by introducing two new notions of generic activities and generic processes and to define specialization/generalization relationships between them. This meta-model is an extension and alteration of this work [4] we published years ago. The remaining of this paper is as follows: Sect. 2 gives the algorithm developed to generate the process variant hierarchy, Sect. 3 reviews the literature and Sect. 4 finally draws some conclusions.

2   Generate a Process Variant Hierarchy

In this section, we show how to apply the transformation sets of generic processes and generic activities and afterwards represent them in a consolidation hierarchy. We develop an algorithm to generate all activity steps of concrete processes derived from generic processes of the Reference process model after applying substitution of each generic activities with respective activity specializations. The steps of this algorithm are as follows: -Firstly, filter (procedure FILTER_STEPS() is left out due to space limitation) only some specific occurrences after applying the sequential order of steps from concrete and generic processes; -Secondly, we generate all steps of concrete processes derived from generic processes after applying direct and non-direct specializations of GAs as described in Algorithm 1. For each direct specialization of generic activities we obtain respective activities from concrete processes as bounded activities. Whereas, for non-direct specializations we use a breadth-first search strategy to explore other activities starting from the specialized activity up to the last activity of a concrete process. -Thirdly, after configuring the genericity levels of the hierarchy by ranking rows according to *lvl*-(GAs absolute level) values. Detail explanations of this algorithm is given in the publication in [5].

Algorithm 1. Derive process variants hierarchy after applying direct and non-direct specializations of GAs from generic processes

Input: $all\_Filtered\_Steps \leftarrow$ FILTER_STEPS()

Output: A multiset $PV\_Hierarchy$ with tuples $\{(act\_id, ga\_id, process\_id, lvl)\}$

    Variables

    $step\_ID = \pi_{StepId}\ (all\_Filtered\_Steps)$

    $process\_ID = \pi_{ProcessId}\ (all\_Filtered\_Steps)$

    $activity\_ID = \pi_{ActivityId}\ (all\_Filtered\_Steps)$

    $lvl = \pi_{lvl}\ (all\_Filtered\_Steps)$ ▷ **step level**

    $isGeneric = \pi_{isGeneric}\ (all\_Filtered\_Steps)$ ▷isGeneric=1 i.e. **step is a GA**

1: **procedure** DERIVE_PV_HIERARCHY()

2:    $PV\_Hierarchy \leftarrow \emptyset$ ▷ multiset of process variants specializations

3:    **foreach** $step\_ID \in all\_Filtered\_Steps$ **do**

4:        **if** $activity\_ID$ *is not null* **then**   ▷ skip control element steps

5:            **if** $isGeneric = true$ **then**   ▷ check if $step\_ID$ is a GA

        ▷ get direct GA's specialization i.e. , an EA or SP

6:            $substituted\_step \leftarrow \pi_{ActivityId}\ \sigma_{GenericActivityId=step\_ID}(a\_is\_spec\_of\_ga)$

7:            $bounded\_step\_a \leftarrow \pi_{S\_Bound.act\_id,\mathbf{step\_ID},\mathbf{process\_ID},lvl}(\rho_{Sub\_S}(substituted\_step) \times$
                                    $\rho_{S\_Bound}(\text{GETBOUNDEDSTEP\_OF\_A}(Sub\_S.ActivityId)))$

        ▷ insert into multiset $PV\_Hierarchy$ with current tuples

8:            $PV\_Hierarchy \leftarrow PV\_Hierarchy\ \cup\ bounded\_step\_a$

        ▷ get indirect GA's specialization using breadth-first-search strategy

9:            $indirect\_step\_a \leftarrow \pi_{BFS.act\_id,\mathbf{step\_ID},\mathbf{process\_ID},lvl}(\rho_{S\_Bound}(bounded\_step\_A) \times$
                                      $\rho_{BFS}(\text{BREADTH\_FIRST\_SEARCH}(S\_Bound.act\_id)))$

        ▷ insert into multiset $PV\_Hierarchy$ with current tuples

10:         $PV\_Hierarchy \leftarrow PV\_Hierarchy\ \cup\ indirect\_step\_a$

        ▷ derive and store process specializations

11:         **get** $concretePId\_of\_a$ for $bounded\_step\_a$

12:         **get** $concretePId\_of\_ind\_a$ for $indirect\_step\_a$

13:         $p\_is\_spec\_of\_gp \leftarrow p\_is\_spec\_of\_gp\ \cup$
                          $\{(concretePId\_of\_a, process\_ID)\}\ \cup$
                          $\{(concretePId\_of\_ind\_a, process\_ID)\}$

14:        **end if**

15:      **end if**

16:    *get* next $step\_ID$ from $all\_Filtered\_Steps$

17: **end foreach**

    ▷ rename $PV\_Hierarchy$ attributes set

18:    $PV\_Hierarchy \leftarrow \pi_{act\_id,\ \rho_{ga\_id/step\_ID},\ \rho_{process\_id/process\_ID},lvl}(PV\_Hierarchy)$

19: **return** $PV\_Hierarchy$

20: **end procedure**

Algorithm 1 as described below generates all steps of concrete processes derived from generic processes after applying direct and non-direct specializations of GAs. As a result, from all these process specializations after substitution operations we derive process variant hierarchy. Accordingly, *Process Variant* dimension in our process warehouse model stores records of this hierarchy. Detailed information the interested reader may find in [3] thesis.

3   Related Work

This section provides an overview of current process-oriented data warehouses approaches to analyse business processes through the most important analysis parameters which are elicited from the generic meta-model of business processes. [2] proposes to derive a generic data warehouse structures from the meta model of the BPMN, whereas [12] proposes a Sequence Warehouse (SeWA) architecture and OLAP tools to analyse data stemming from workflow logs but a conceptual model for DW is missing. Approaches based on goal-oriented methodology for requirement analysis in order to design a data warehouse were proposed by [11,16,20]. A multidimensional data modelling for business process analysis was proposed by [8,14,15,17]. During our research work we have found a number of relational and multidimensional data warehouse design used for process mining analysis as well. According to approach in [1], process cubes notion is presented to organize events and mined process models using different dimensions. Authors in [19] introduced an event cube as a basis for process discovery and analysis. A framework is further developed to realize a process cube allowing for the comparison of event data in [13]. A hierarchy level was defined only in the time dimension. For instance, multidimensional process mining can be used to analyse the different versions of a sales process, where each version can be defined according to different dimensions such as location or time, and then the different results can be compared as proposed in [6]. Furthermore, authors in [21] partition event logs into groups of cases called sublogs with homogeneous features in a dynamic and flexible way, in order to manage comparisons between models. Whereas, authors in [10] proposed an Abstract Argumentation Framework (AAF) to support the "high-level" analysis that business analysts are used to reason.

4   Conclusions

This paper proposed an algorithm that consolidates a family of process variants into a hierarchy of a process warehouse solution to efficiently and effectively analyse them. Current business process management systems and traditional process warehouses lack on adequately abstracting and consolidating all variants into one generic process model, to provide the possibility to distinguish and compare among different parts of different variants. As a summary, based on the consumption of PW in many business intelligence development and solutions, a framework that allows process variants to be efficiently analysed can significantly improve the state-of-art.

References

1. van der Aalst, W.M.P.: Process cubes: slicing, dicing, rolling up and drilling down event data for process mining. In: Song, M., Wynn, M.T., Liu, J. (eds.) AP-BPM 2013. LNBIP, vol. 159, pp. 1–22. Springer, Cham (2013). https://doi.org/10.1007/978-3-319-02922-1_1

2. Benker, T.: A generic process data warehouse schema for BPMN workflows. In: Abramowicz, W., Alt, R., Franczyk, B. (eds.) BIS 2016. LNBIP, vol. 255, pp. 222–234. Springer, Cham (2016). https://doi.org/10.1007/978-3-319-39426-8_18

3. Berberi, L.: Analysis of Process Variants with a Process Warehouse Approach. Ph.D. thesis (2021)

4. Berberi, L., Eder, J., Koncilia, C.: A process warehouse model capturing process variants. EMISA J. (2018)

5. Berberi, L.: A process warehouse for process variants analysis, p. 17. KITopen (2022). https://doi.org/10.5445/IR/1000146946

6. Bolt, A., van der Aalst, W.M.P.: Multidimensional process mining using process cubes. In: Gaaloul, K., Schmidt, R., Nurcan, S., Guerreiro, S., Ma, Q. (eds.) CAISE 2015. LNBIP, vol. 214, pp. 102–116. Springer, Cham (2015). https://doi.org/10.1007/978-3-319-19237-6_7

7. BPMN(Spec.): about the business process model and notation specification version 2.0. Technical report, OMG (2011). https://www.omg.org/spec/BPMN/2.0/PDF

8. Casati, F., Castellanos, M., Dayal, U., Salazar, N.: A generic solution for warehousing business process data. VLDB Endowment (2007)

9. Eder, J., Olivotto, G.E., Gruber, W.: A data warehouse for workflow logs. In: Han, Y., Tai, S., Wikarski, D. (eds.) EDCIS 2002. LNCS, vol. 2480, pp. 1–15. Springer, Heidelberg (2002). https://doi.org/10.1007/3-540-45785-2_1

10. Fazzinga, B., Flesca, S., Furfaro, F., Pontieri, L.: Process mining meets argumentation: explainable interpretations of low-level event logs via abstract argumentation (2022)

11. Giorgini, P., Rizzi, S., Garzetti, M.: Goal-oriented requirement analysis for data warehouse design. ACM (2005)

12. Koncilia, C., Pichler, H., Wrembel, R.: A generic data warehouse architecture for analyzing workflow logs. In: Morzy, T., Valduriez, P., Bellatreche, L. (eds.) ADBIS 2015. LNCS, vol. 9282, pp. 106–119. Springer, Cham (2015). https://doi.org/10.1007/978-3-319-23135-8_8

13. Mamaliga, T.: Realizing a Process Cube Allowing for the Comparison of Event Data. Master's thesis, TU Eindhoven (2013)

14. Mansmann, S., Neumuth, T., Scholl, M.H.: OLAP technology for business process intelligence: challenges and solutions. In: Song, I.Y., Eder, J., Nguyen, T.M. (eds.) DaWaK 2007. LNCS, vol. 4654, pp. 111–122. Springer, Heidelberg (2007). https://doi.org/10.1007/978-3-540-74553-2_11

15. Neumuth, T., Mansmann, S., Scholl, M.H., Burgert, O.: Data warehousing technology for surgical workflow analysis. IEEE (2008)

16. Niedrite, L., Solodovnikova, D., Treimanis, M., Niedritis, A.: Goal-driven design of a data warehouse-based business process analysis system. In: AIKED 2007 (2007)

17. Pau, K.C., Si, Y.W., Dumas, M.: Data warehouse model for audit trail analysis in workflows. IEEE (2007)

18. Polyvyanyy, A., Ouyang, C., Barros, A., van der Aalst, W.M.: Process querying: Enabling business intelligence through query-based process analytics. Decis. Support Syst. (2017)

19. Ribeiro, J.T.S., Weijters, A.J.M.M.: Event cube: another perspective on business processes. In: Meersman, R., et al. (eds.) OTM 2011. LNCS, vol. 7044, pp. 274–283. Springer, Heidelberg (2011). https://doi.org/10.1007/978-3-642-25109-2_18

20. Shahzad, K., Zdravkovic, J.: Process warehouses in practice: a goal-driven method for business process analysis. J. Softw. Evol. Process. **24**, 321–339 (2012)
21. Vogelgesang, T., Appelrath, H.-J.: A relational data warehouse for multidimensional process mining. In: Ceravolo, P., Rinderle-Ma, S. (eds.) SIMPDA 2015. LNBIP, vol. 244, pp. 155–184. Springer, Cham (2017). https://doi.org/10.1007/978-3-319-53435-0_8

Feature Selection Algorithms

Unsupervised Features Ranking via Coalitional Game Theory for Categorical Data

Chiara Balestra[1(✉)], Florian Huber[3], Andreas Mayr[2], and Emmanuel Müller[1]

[1] TU Dortmund University, Dortmund, Germany
chiara.balestra@cs.tu-dortmund.de
[2] University Hospital of Bonn, Bonn, Germany
[3] University of Bonn, Bonn, Germany

Abstract. Not all real-world data are labeled, and when labels are not available, it is often costly to obtain them. Moreover, as many algorithms suffer from the curse of dimensionality, reducing the features in the data to a smaller set is often of great utility. Unsupervised feature selection aims to reduce the number of features, often using feature importance scores to quantify the relevancy of single features to the task at hand. These scores can be based only on the distribution of variables and the quantification of their interactions. The previous literature, mainly investigating anomaly detection and clusters, fails to address the redundancy-elimination issue. We propose an evaluation of correlations among features to compute feature importance scores representing the contribution of single features in explaining the dataset's structure.

Based on Coalitional Game Theory, our feature importance scores include a notion of redundancy awareness making them a tool to achieve redundancy-free feature selection. We show that the deriving features' selection outperforms competing methods in lowering the redundancy rate while maximizing the information contained in the data. We also introduce an approximated version of the algorithm to reduce the complexity of Shapley values' computations.

Keywords: Unsupervised learning · Features importance · Redundancy reduction

1 Introduction

In machine learning, both feature selection methods and reduction of dimensionality are often performed to increase interpretability and to reduce computational complexity. As an example, for unsupervised applications such as clustering [5] or anomaly detection [14], the curse of dimensionality poses a major challenge. Unsupervised feature selection enables the detection of data patterns, as well as the description of these patterns using a concise set of relevant features [20,24]. The corresponding methods are mostly based on the analysis of multivariate data distributions, pairwise correlations, higher-order interactions among features, or pseudo-labels. The use of such complex

C. Balestra—This research was supported by the research training group *Dataninja* (Trustworthy AI for Seamless Problem Solving: Next Generation Intelligence Joins Robust Data Analysis) funded by the German federal state of North Rhine-Westphalia.

R. Wrembel et al. (Eds.): DaWaK 2022, LNCS 13428, pp. 97–111, 2022.
https://doi.org/10.1007/978-3-031-12670-3_9

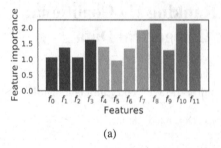
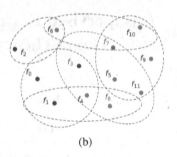

(a)   (b)

Fig. 1. (a) Unsupervised Shapley values-based feature importance scores. (b) Shapley values consider interactions within all possible subsets of features f_is. In both figures, correlated subsets of features are color-coded and features selected by the proposed algorithm are marked in red. (Color figure online)

measures implies that both the selection as well as the interpretation of why some features have been selected is challenging. On the one hand, selection requires basic measures to quantify the interaction within a set of features [5,14,16]. On the other hand, interpretation of higher-order interactions is non-straight-forward and requires the decomposition of complex non-linear, higher-order, and multivariate measures to feature importance scores.

In different application domains, raising understanding over the mechanisms of underlying machine learning techniques has become a crescent necessity. Assigning scores to features based on their contributions to the machine learning procedure plays a decisive role to this end. Feature importance scores are prevalent in supervised learning, e.g., random forests. At the same time, for unsupervised tasks, the literature is limited either to traditional scores [20,24] not sensitive to higher-order interactions, or the scores are not easily interpretable higher-order correlation measures.

We propose new unsupervised feature importance scores decomposing the information contained in the data using axiomatic game-theoretic properties. In particular, Shapley values enable us to consider the interactions present in each possible subset of features (Fig. 1(b)) and to assign importance scores to the single features accordingly. Our approach consists of two steps. In the first step, we introduce a game-theoretic solution to decompose the information contained in the dataset and assign importance scores to the single features. The scores obtained consider complex higher-order feature interactions, can be based on different correlation measures, and do not rely on specific notions of clustering or anomalies. In particular, we use Shapley values [18] to get the feature importance scores where features explaining the most information on the overall dataset obtain a higher score. In the second step, we take care of a mechanism to reduce the redundancy among features. To this end, feature importance scores are penalized through an information-theoretic measure of correlation to yield a redundancy-free feature selection. Figure 1(a) displays how correlated features are ranked similarly before applying the redundancy elimination step and how our method is capable of avoiding selecting highly correlated features. In the experimental results, we show that our ranking is achieving a redundancy free-ranking; the redundancy rate of the selected features is kept low both in synthetic and real datasets.

As a final remark, the scores flexibly rely on different correlation measures and are not bound to any clustering or anomaly detection goals. We choose to present the

Table 1. Summary table of the competing methods and this paper.

	versatile quality notion	feature ordering	iterative selection	redundancy awareness	higher-order interactions
UDFS [25]	✗	✓	✗	✓	✓
MCFS [2]	✗	✓	✗	✓	✗
NDFS [11]	✗	✓	✗	✓	✗
SPEC [26]	✓	✓	✗	✗	✗
LS [8]	✓	✓	✗	✗	✗
PFA [12]	✓	✗	✗	✓	✗
FSFC [27]	✗	✗	✓	✓	✗
this paper	✓	✓	✓	✓	✓

results obtained using the total correlation; hence, the presented experimental results are limited to discrete and categorical data. The procedure can be extended to mixed datasets replacing the total correlation or applying discretization on continuous data.

2 Related Work

Dimensionality reduction helps avoid the curse of dimensionality and increases the interpretability of data and machine learning techniques. Different methods analyze the relationship among features, the class label, and the correlation among variables [23] and get feature importance scores in order to allow for a more aware use of machine learning by non-experts. Those scores are often not aware of correlations among variables, thus leading to a necessary integration of a redundancy awareness concept [19].

In 2007 game theory found application in supervised feature selection [6,15] where the value function was defined as the accuracy or the generalization error of the trained model; to the best of our knowledge, the approaches proposed in the recent years are limited to labeled data. A recent paper [17] underlined how Shapley values spread through machine learning; in particular, they appear in several techniques to increase the overall interpretability of black-box models [13,21] and new insights on Shapley values and their applications continue appearing in the literature [4].

As a downside, Shapley values are well known to be computationally expensive. Several approximations found place in the literature, e.g., [1,3,22] among others; the first attempt of a comprehensive survey of Shapley values' approximations is represented by Rozemberczki et al. [17]. To reduce the computational run-time, we implement Castro et al. [3] approximation, i.e., the most common Shapley values' approximation non relying on additional assumptions on the players.

As a parallel area of research, in recent years, unsupervised feature selection methods have raised strong interest in the community [10,20]. We selected a representative sample within the vast number of unsupervised feature selection methods to compare the performance of our approach. Among them, UDFS [25] creates pseudo-labels to perform the feature selection in unlabelled data; MCFS and NDFS [2,11] concentrate on keeping the clustering structure. LS [8] selects features by their local preserving power. PFA [12] tries to eliminate the downside of PCA while keeping the information within the data. Most of these algorithms tend to select features as a by-product of retaining a clustering structure in the data. Finally, FSFC [27] is meant to select only non-redundant variables using a new definition of distance in the k-nearest neighbors. Table 1 illustrates a summary of the properties of the various methods analyzed in comparison with our paper.

3   Feature Importance Measures

Consider a N-dimensional dataset containing D instances. We interpret each dimension as the realization set of a random variable, refer to the set of variables as $\mathcal{F} = \{X_1, \ldots, X_N\}$ and to each dimension X_i as ith feature or variable. Feature selection methods often internally assign to subsets of features an importance score and output the subset maximizing the mentioned score. We propose to rank features considering their average contribution to all the possible subsets of features. The higher the average contribution of a feature is, the more convenient it is to keep it within the selected features. Additionally, we will also introduce redundancy awareness in these scores.

Given a function that assigns a value to each subset of features, assessing the *importance* of single features is not trivial as each feature belongs to 2^{N-1} subsets of features. In unsupervised contexts, we can assess the usefulness of a set of features measuring correlations or clustering properties. Throughout the manuscript, we stick to a value function that captures the maximal *information* contained in the data. Following this choice, the approach presented is restricted to categorical tabular data. We compute feature importance scores and obtain a ranking prioritizing features highly correlated with the rest of the dataset.

3.1   Feature Importance Score

We obtain feature importance scores using coalitional game theory. Each game is fully represented by the set of players $\mathcal{F}$ and a set function v that maps each subset $\mathcal{A} \subseteq \mathcal{F}$ to $v(\mathcal{A}) \in \mathbb{R}$. v is referred to as *value function* [18] and satisfies the following properties

1. $v(\emptyset) = 0$,
2. $v(\mathcal{A}) \geq 0$ for any $\mathcal{A} \subseteq \mathcal{F}$, and
3. $v(\mathcal{A}) \leq v(\mathcal{B})$ for any $\mathcal{A}, \mathcal{B} \subseteq \mathcal{F}$ such that $\mathcal{A} \subseteq \mathcal{B}$.

Working with unlabelled data, we can not rely on ground truth labels. Hence, we define value functions relying on intrinsic properties of the dataset; we opt for a value function measuring the independence of the features in $\mathcal{A} \subseteq \mathcal{F}$. One possible initialization for v is the *total correlation* of $\mathcal{A}$.

Definition 1. *The* total correlation C *of a set of variables* $\mathcal{A} \subseteq \mathcal{F}$ *is defined as*

$$C(\mathcal{A}) = \sum_{X \in \mathcal{A}} H(X) - H(\mathcal{A}). \tag{1}$$

$H(\mathcal{A})$ *is the Shannon entropy of the subset of discrete random variables* $\mathcal{A}$, *i.e.,*

$$H(\mathcal{A}) = -\sum_{\vec{x} \in \mathcal{A}} p_{\mathcal{A}}(\vec{x}) \log p_{\mathcal{A}}(\vec{x}) \tag{2}$$

where $p_{\mathcal{A}}(\cdot)$ *is the joint probability mass function of* $\mathcal{A}$.
$H(X)$ *is the Shannon entropy of* X, *i.e.,* $H(X) = -\sum_{x \in X} p_X(x) \log p_X(x)$.

We choose the total correlation as it satisfies properties (2) and (3), it has an intuitive meaning and can be easily extended such that it satisfies property (1).

Shannon entropy [7] measures the uncertainty contained in a random variable X considering how uniform data are distributed: its value is close to zero when its probability mass function p_X is highly skewed while, as the distribution approaches a uniform distribution, its value increases. Moreover, the Shannon entropy is a monotone non-negative function and can be extended such that $H(\emptyset) = 0$. We assume that all features in $\mathcal{F}$ are discrete as the extension of Shannon entropy to continuous variables is not monotone [9]. As a consequence of Shannon entropy's properties, the total correlation $C(\mathcal{A})$ is close to zero if the variables in $\mathcal{A}$ are independent, and it increases when they are correlated. To study the impact of adding a feature Y to $\mathcal{A} \subseteq \mathcal{F}$, we compute the value function of the incremented subset $v(\mathcal{A} \cup Y)$ and compare it with $v(\mathcal{A})$: The difference $v(\mathcal{A} \cup Y) - v(\mathcal{A}) = H(\mathcal{A}) + H(Y) - H(\mathcal{A} \cup Y)$ is non-negative and measures how much $\mathcal{A}$ and Y are correlated. We refer to $H(\mathcal{A}) + H(Y) - H(\mathcal{A} \cup Y)$ as *marginal contribution of Y to $\mathcal{A}$*. If $\mathcal{A}$ and Y are independent, then the marginal contribution of Y to $\mathcal{A}$ equals zero. Vice versa, the marginal contribution grows the stronger the correlation between Y and $\mathcal{A}$ is. As importance score, we assign to X_i the average of its marginal contributions and we refer to it as $\phi(X_i)$, i.e.,

$$\phi(X_i) = \sum_{\mathcal{A} \subseteq \mathcal{F} \setminus X_i} \frac{1}{N\binom{N-1}{|\mathcal{A}|}} [H(\mathcal{A}) + H(X_i) - H(\mathcal{A} \cup X_i)] \qquad (3)$$

corresponding to the *Shapley value* of the player X_i in the game $(\mathcal{F}, v)$ when v is the total correlation. The general definition of Shapley values reads [18]:

Definition 2. *Given a coalitional game $(\mathcal{F}, v)$ and a player $X_i \in \mathcal{F}$, the Shapley value of X_i is defined by*

$$\phi_v(X_i) = \sum_{\mathcal{A} \subseteq \mathcal{F} \setminus X_i} \frac{1}{N\binom{N-1}{|\mathcal{A}|}} [v(\mathcal{A} \cup X_i) - v(\mathcal{A})].$$

It can be proven that the Shapley value is the only function that satisfies the *Pareto optimality*, i.e., $\sum_{X_i \in \mathcal{F}} \phi_v(X_i) = v(\mathcal{F})$, the dummy, the symmetry and additive properties [18]. Moreover, Shapley values represent a fair assignment of resources to players based on their contributions to the game. We use the scores $\phi(X_i)$ to rank the features in the dataset $\mathcal{F}$. However, Shapley values do not consider redundancies, and linearly dependent features obtain equal Shapley values.

3.2   Importance Scores of Low Correlated Features

We use a dataset with three sets of correlated features (color-coded in Fig. 1(a)), and we aim to select features from subsets with different colors; however, as we have already underlined, correlated features are characterized by similar Shapley values. In particular, the three highest Shapley values are obtained by correlated features in the blue-colored set. Before addressing the problem of redundancy-awareness inclusion in Shapley values, we show that the Shapley values rank features that are not correlated with the rest of the dataset in low positions.

Algorithm 1. SVFS

```
1: procedure SVFS(F, ε)
2:     S = ∅
3:     while F ≠ ∅ do
4:         while X ∈ F do
5:             if H(X) + H(S) − H(S, X) > ε then
6:                 F = F \ X
7:             else
8:                 F = F
9:                 S = S ∪ arg max_{X∈F}{φ(X)}
10:                F = F \ S
       return S
```

Algorithm 2. SVFR

```
1: procedure SVFR(F)
2:     S = arg max_{X∈F}{φ(X)}
3:     ordered = [ ]
4:     ordered[0] = arg max_{X∈F}{φ(X)},   j = 1
5:     F = F \ S
6:     while F ≠ ∅ && j < N do
7:         for X ∈ F do
8:             rk(X) = φ(X) − H(X) − H(S) + H(S, X)
9:         ordered[j] = arg max_{X∈F}{rk(X)}
10:        S = S ∪ arg max_{X∈F}{rk(X)}
11:        F = F \ S
12:        j + +
     return ordered
```

Theorem 1. *Given a subset of features $\mathcal{B} \subset \mathcal{F}$ that satisfies the following properties*

1. for all $X_j \notin \mathcal{B}$ and for all $\mathcal{A} \subseteq \mathcal{F} \setminus \{X_j\}$, $H(\mathcal{A}) + H(X_j) = H(\mathcal{A} \cup X_j)$
2. for all $X_i \in \mathcal{B}$ and for all $\mathcal{A} \subseteq \mathcal{F} \setminus \{X_i\}$, $H(\mathcal{A}) + H(X_i) \geq H(X_i \cup \mathcal{A})$

then $\phi(X_i) \geq \phi(X_j)$ for all $X_i \in \mathcal{B}$ and $X_j \notin \mathcal{B}$.

Proof. From (1) we know that, since the marginal contribution of $X_j \notin \mathcal{B}$ to any $\mathcal{A} \subseteq \mathcal{F} \setminus \{X_j\}$ is equal to zero, $\phi(X_j) = \sum_{\mathcal{A} \subseteq \mathcal{F} \setminus \{X_j\}} \frac{1}{N\binom{N-1}{|\mathcal{A}|}} \cdot 0 = 0$.

For any $X_i \in \mathcal{F}$ and $\mathcal{A} \subseteq \mathcal{F}$, we know that $H(\mathcal{A} \cup X_i) \leq H(\mathcal{A}) + H(X_i)$ from Shannon entropy's properties [7]. Hence, all marginal contributions are non-negative. Hence, $\phi(X_i) \geq 0 = \phi(X_j)$ for all $X_i \in \mathcal{B}$ and $X_j \notin \mathcal{B}$.

This concludes the proof.

Thus with total correlation as value function, Shapley values are non-negative and equal zero if and only if the feature is non-correlated with any subset of features. Moreover, features highly correlated with other subsets of features get high Shapley values.

4   Redundancy Removal

We address the challenge of adding redundancy awareness to Shapley values. For this purpose, we develop a pruning criteria based on the total correlation and greedily rank features to get a redundancy-free ranking of features while still looking for features with high Shapley values. Feature selection based on this ranking selects the variables ranked first by Shapley values which show little dependencies.

We propose two algorithms. The Shapley Value Feature Selection (SVFS) needs a parameter ϵ representing the correlation among features that we are willing to accept; hence, SVFS requires some expert knowledge on the dataset to specify the parameter ϵ in an opportune interval. The Shapley Value Feature Ranking (SVFR) works automatically with an included notion of redundancy. We show that the two algorithms lead to consistent results in Sect. 6.5. At each step, both algorithms select the highest-ranked feature among the ones left.

We use a total correlation-based punishment; In particular, $H(\mathcal{A}) + H(X) - H(\mathcal{A} \cup X) \geq 0$ represents the strength of the correlation among X and $\mathcal{A}$ and it is equal to zero if and only if X and $\mathcal{A}$ are independent.

SVFS's inputs are the set of unordered features $\mathcal{F}$ and the parameter $\epsilon > 0$; ϵ plays the role of a stopping criterion and represents the maximum correlation that we are willing to accept within the set of selected features. Whenever ϵ is high, we end up with the ordering given by Shapley values alone; instead, for $\epsilon \approx 0$ the criterion can lead to the selection of the only features which are uncorrelated with the first one. The optimal range of ϵ highly depends on the dataset. We show that SVFS is robust w.r.t. the choice of ϵ. At each iteration, SVFS excludes from the ranking the features Xs that are correlated with the already ranked features $\mathcal{S} \subseteq \mathcal{F}$ more than ϵ, i.e., $H(X) + H(\mathcal{S}) - H(\mathcal{S}, X) > \epsilon$, computes the Shapley values of all remaining features X and adds to $\mathcal{S}$ the feature whose Shapley value is the highest. When there are no features left, it stops and returns $\mathcal{S}$.

SVFR takes as an input $\mathcal{F}$ and outputs a feature ranking without the need of any additional parameter. The ranking is aware of correlations as each of the Shapley values $\phi(X_i)$ is penalized using the correlation measure $H(X_i) + H(\mathcal{S}) - H(X_i \cup \mathcal{S})$ where $\mathcal{S}$ is the set of already ranked features, and X_i is a new feature to be ranked. This algorithm provides a complete ranking of features and can be prematurely stopped including an upper bound of features we are willing to rank. The absence of the additional parameter ϵ is the main advantage of SVFR over SVFS.

5   Scalable Algorithms

The size of $\mathcal{P}(\mathcal{F})$ being exponential in N, computing Shapley values involves 2^N evaluations of the value function. We use approximated Shapley values to obtain scalable versions of SVFR and SVFS. We implement three versions of the algorithms that differ only in the computations of Shapley values used:

- *full algorithm*: it uses the full computation of the Shapley values
- *bounded algorithm*: consider only subsets up to size k fixed to compute the Shapley values

– *sampled algorithm*: it uses the approximation proposed by Castro et al. [3] based on n random sampled subsets of features.

The time complexity for the sampled algorithm is $\mathcal{O}(D \cdot n)$, for the bounded algorithm is $\mathcal{O}(D \cdot N^k)$ while for the full algorithm is $\mathcal{O}(D \cdot 2^N)$ where N is the number of features and D the number of samples in the dataset.

6 Experiments

We show that our feature ranking method outperforms competing representative feature selection methods in terms of redundancy reduction. Metrics such as NMI, ACC, and redundancy rate are often used in the previous literature to evaluate unsupervised feature selection methods. NMI and ACC focus on the cluster structure in the data; therefore, as clustering is not the goal of our approach, we compare it with the competing methods using the redundancy rate. The redundancy rate of $\mathcal{S} \subseteq \mathcal{F}$ is defined in terms of pairwise Pearson correlations, i.e.,

$$\text{Red}(\mathcal{S}) = \frac{1}{2m(m-1)} \sum_{X,Y \in \mathcal{S}, X \neq Y} \rho_{X,Y} \qquad (4)$$

where $\rho_{X,Y} \in [0,1]$ is the Pearson correlation of features X and Y. It represents the averaged correlation among the pairs of features in $\mathcal{S}$ and varies in the interval $[0,1]$: a $\text{Red}(\mathcal{S})$ close to 1 shows that many selected features in $\mathcal{S}$ are strongly correlated while a value close to zero indicates that $\mathcal{S}$ contains little redundancy. In the experiments, we use the *redundancy rate* as evaluation criteria re-scaling it to the interval $[0, 100]$ via the maximum pair-wise correlation to facilitate the comparison among different datasets.

6.1   Datasets and Competing Methods

We show a comparison against *SPEC, MCFS, UDFS, NDFS, PFA, LS* and FSFC [2, 8, 11, 12, 25–27].

We use various synthetic and publicly available datasets: the *Breast Cancer dataset*, the *Big Five Personalities Test dataset*[1] and the *FIFA dataset*[2]. The datasets that we use throughout the paper are all categorical or discrete. We consider subsets of the full dataset in order to apply the full versions of the algorithms and investigate the performance of the approximations of SVFR and SVFS at the end of the section.

6.2   Redundancy Awareness

We compare the feature selection results of our algorithm against the competitors by evaluating the redundancy rate in Table 2. For the FIFA dataset, we select 15 features

---

[1] The first 50 features in the Big Five dataset are the categorical answers to the personality test's questions and are divided into 5 personalities' traits (10 questions for each personality trait). To apply the full algorithm, we select questions from different personalities and restrict to 10000 instances.

[2] We restrict to the 5000 highest-rated players by the overall attribute.

Table 2. Redundancy rate of the sets of three selected features using the competing algorithms and SVFR (highlighted in green color in the table) on different datasets. The lowest rates are represented in bold characters.

	Breast Cancer	B5_balanced	B5_unbalanced	FIFA	Synthetic
NDFS	36.30	22.11	20.75	18.97	**1.49**
MCFS	20.26	23.59	18.79	20.63	3.74
UDFS	33.59	28.13	35.18	57.73	4.06
SPEC	13.89	39.09	21.46	42.14	29.4
LS	7.05	28.83	58.25	48.28	100.00
PFA	**5.10**	23.22	34.28	57.42	35.84
FSFC	8.74	22.64	20.99	36.45	2.12
SVFR	6.68	**15.65**	**18.02**	**14.79**	1.51

from the entire data which characterize the *agility*, *attacking* and *defending* skills of the football players; we keep the whole datasets for Breast Cancer and synthetic data; in the case of the Big Five Personality Traits dataset, we select respectively 5 questions from three different personality traits for the balanced dataset and 9 features from one trait and 3 from other two personality traits in the case of the unbalanced dataset. In order to avoid bias towards the random selection of personality traits and features in the Big Five data, we average the redundancy rate over 30 trials on randomly selected personalities and variables both in the case of the balanced and unbalanced setup.

In each column, bold characters highlight the lowest redundancy rate. We use SVFR for ranking the features and select the three highest-ranked features. We consequently specified the parameters of the competing methods in order to get a selection of features as close to three features as possible. For FCFS we set $k = 4$ for BC dataset, $k = 8$ for FIFA dataset, $k = 8$ for the synthetic data and for *Big Five* dataset we use different k at each re-run such that the number of selected variables varies between 2 and 5 and then we average the redundancy rates; for NDFS, MCFS, UDFS and LS we used $k = 5$ (k being the number of clusters in the data); for the other competitors, we specify the number of features to be selected. Table 2 illustrates that SVFR outperforms the competing methods in nearly all the cases. In particular, while SVFR achieves low redundancy rates in all datasets, the competing algorithms show big differences in performance in the various datasets. On the Breast Cancer data and the synthetic dataset respectively, PFA and NDFS slightly outperform SVFR. However, they do not keep an average low redundancy rate on the other datasets. For reproducibility, we make the code publicly available[3].

6.3 Relevance of Unsupervised Feature Selection and Effectiveness

In Fig. 2(a), each plot corresponds to a different subset of features of the Big Five dataset, i.e., 10 features selected from three different personality traits. Running SVFS with $\epsilon = 0.3$ we detect correlated features and avoids selecting them together as shown in the plots. Using the scaled versions of our algorithms from Sect. 5 we can extend the approach towards the complete Big Five dataset.

---

[3] https://github.com/chiarabales/unsupervised_sv.

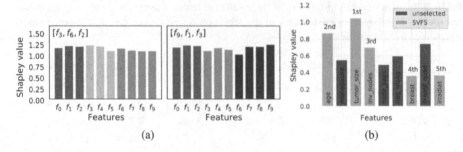

Fig. 2. (a) Barplot of Shapley values and respective feature selections by SVFS ($\epsilon = 0.3$) in Big Five dataset restricted to 10 features. Different personalities traits are color-coded in each plot. (b) Barplot of Shapley values for Breast cancer data; in green, the ordering of features' selection by SVFS when $\epsilon = 0.5$. (Color figure online)

Table 3. Orderings of selection given by SVFS for various ϵ and first 8 ranked features by SVFR. Features are color-coded in order to simplify the visualization.

	Big Five	Synthetic Data	Breast Cancer
$\epsilon = 0.2$	$[11, 0, 5]$	$[8, 7, 0]$	$[2, 0, 8]$
$\epsilon = 0.3$	$[11, 0, 10]$	$[8, 7, 2]$	$[2, 0, 4, 6]$
$\epsilon = 0.4$	$[11, 0, 14]$	$[8, 7, 3]$	$[2, 0, 4, 8, 6]$
$\epsilon = 0.5$	$[11, 0, 14, 9]$	$[8, 7, 3]$	$[2, 0, 3, 8, 6]$
$\epsilon = 0.6$	$[11, 0, 14, 5]$	$[8, 7, 3]$	$[2, 0, 3, 8, 6]$
$\epsilon = 0.7$	$[11, 0, 14, 13]$	$[8, 7, 3]$	$[2, 0, 3, 4, 8, 6]$
$\epsilon = 0.8$	$[11, 0, 14, 13]$	$[8, 7, 3, 0]$	$[2, 0, 3, 4, 5, 8, 6]$
SVFR	$[11, 0, 5, 10, 12, 8, 6, 2]$	$[8, 7, 3, 0, 6, 5, 2, 10]$	$[2, 0, 4, 6, 8, 5, 1, 3]$

Figure 1(a) represents the Shapley values of features in a 12 dimensional synthetic dataset where subsets of correlated features are color-coded. We measure the ability of the algorithm in selecting features from different subsets of correlated features; SVFS selects one feature from each subset of correlated features. In particular, when $\epsilon = 1$, SVFS achieves this goal by selecting $\{f_8, f_7, f_3\}$ while the ranking given by the Shapley values alone is $\{f_8, f_{10}, f_{11}\}$ which belong to the same subset of correlated features. This nicely underlines the inability of Shapley values to detect correlations and the necessity of integrating correlation-awareness to perform a feature selection.

Our unsupervised feature selection allow to construct more efficient psychological tests avoiding redundancies and reducing the number of questions that need to be answered without losing too much information.

6.4   Interpretation of Feature Ranking

We apply SVFS when $\epsilon = 0.5$ to the Breast Cancer dataset. In Fig. 2(b), the resulting Shapley values and the ordering of selected features are displayed. The selection

Table 4. $recall@k$ for $k \in \{1,3,5\}$ comparing a random ranking and the rankings given by SVFR using the sampled and bounded algorithms to the full SVFR ranking. We show results for FIFA and Big Five datasets restricted to 15 features randomly chosen. Bold text highlights the best approximation.

		$k = 1$	$k = 3$	$k = 5$
BIG5	Random	0.04	0.19	0.33
	Sampled	0.04	0.37	0.49
	Bounded	**0.08**	**0.56**	**0.55**
FIFA	Random	0.06	0.24	0.35
	Sampled	0.00	0.33	0.40
	Bounded	**1.00**	**0.67**	**0.80**

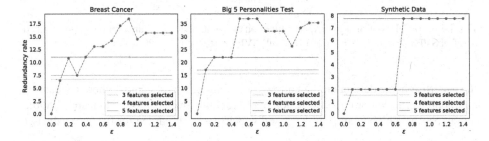

Fig. 3. Redundancy rates of the selected features' sets as a function of ϵ for SVFS (bullets points connected by the dashed line) and for $3, 4$, and 5 selected features when using SVFR.

resulting from SVFS shows a low redundancy rate while the selected features (e.g., the size of the tumour, age, and the number of involved lymph nodes) are clearly in line with domain knowledge on risk factors for disease progression (label). Furthermore, the comparison with the ranking without redundancy awareness nicely highlights the importance of our approach to avoid redundancies when possible.

6.5   Comparison Among the Proposed Algorithms

In Fig. 3, we plot a comparison among SVFS and SVFR w.r.t. the redundancy rate on three datasets with different values of ϵ. As benchmarks, we use for SVFR the selection of 3, 4 and 5 features respectively while for SVFS, ϵ varies in the interval $[0, 1.4]$ with steps of size 0.1.

Using the number of features as a stopping criterion in SVFR would produce consistent results to SVFS: as an example, using the breast cancer data the ranking given by SVFR, i.e., $[2, 0, 4, 6, 8, 5, 1, 3]$, is consistent with the selection given by SVFS respectively using $\epsilon = 0.2$ and $\epsilon = 0.6$, i.e., $[2, 0, 8]$ and $[2, 0, 3, 8, 6]$. Table 3 shows a full comparison among the SVFR and SVFS on three different representative datasets. We recommend applying SVFS when no previous knowledge of the data is available and it is hard to establish an optimal range for ϵ. Vice versa, one could apply SVFR when

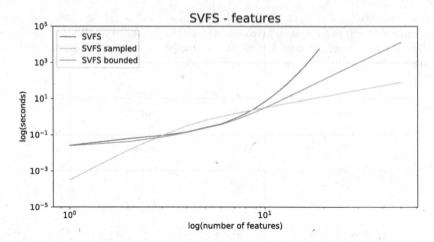

Fig. 4. Log-log plots of the run-time as a function of the number of features for the approximated and full SVFS ($\epsilon = 0.5$, $D = 1000$). The full SVFS is stopped with 20 features.

the expertise in the dataset domain allows determining a reasonable number of features as stopping criterion or the observation of the ranking given can provide insights to the non-expert on which features to keep and which can be discarded for further analysis.

6.6 Run-Time Analysis

As a consequence of the full computation of Shapley values, the run-time of SVFR and SVFS increases exponentially with the number of features as shown by Fig. 4. Using the approximated algorithms, this growth turns out to be slower. In particular, when using the sampled algorithm, the run-time increases only linearly with the number of features while the growth of the bounded algorithm's run-time is polynomial in the number of features. In the additional material, we show the log-log plot of the run-time for increased number of samples in the dataset. For each algorithm, we use random subsets of the Big Five dataset and average over 10 trails.

We further compare the rankings of the approximated and full algorithms using the $recall@k$ metric interpreting rankings of the full version of SVFR as ground truth. We use the Big Five dataset, randomly selecting 5 questions from 3 different personalities and average the scores over 100 trails (see Table 4). Overall, the results for the approximated algorithms clearly outperform random ordering - but still deviate often from the full versions. It is worth to note that the bounded algorithm using subsets up to size 5 performs better than the sampled version.

7 Conclusions

In the paper, we develop a new method to assess feature importance scores in unsupervised learning, bridging the gap between unsupervised feature selection and cooperative game theory. We integrate Shapley values with redundancy awareness making use of an entropy-based function to get feature importance scores.

We present two algorithms: SVFS implements feature selection using a redundancy aware criterion while SVFR assigns a ranking to each feature while being aware of correlations with previously ranked features. We show how the results of the two algorithms are consistent and state-of-the-art regarding their application. Our feature selection methods outperform previously proposed algorithms w.r.t. the redundancy rate. We additionally introduce approximated versions of the algorithms that are scalable to higher dimensions.

Additional Material

(See Fig. 5 and Table 5).

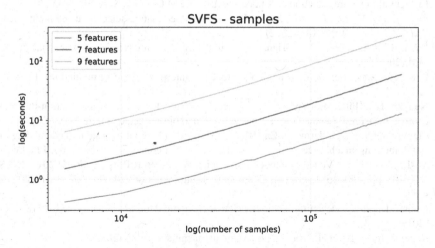

Fig. 5. Log-log plot of the run-time for the full SVFS with $\epsilon = 0.5$ as a function of the number of the samples D and fixed number of features.

Table 5. Summary of the datasets' structures.

	Features	Samples
Breast Cancer dataset	9	286
Big Five dataset	50	1013558
FIFA20 dataset	46	15257
Synthetic dataset	12	10000

References

1. Burgess, M.A., Chapman, A.C.: Approximating the Shapley value using stratified empirical Bernstein sampling. In: IJCAI (2021)
2. Cai, D., Zhang, C., He, X.: Unsupervised feature selection for multi-cluster data. In: KDD (2010)
3. Castro, J., Gómez, D., Tejada, J.: Polynomial calculation of the Shapley value based on sampling. Comput. Oper. Res. **36**, 1726–1730 (2009)
4. Catav, A., et al.: Marginal contribution feature importance - an axiomatic approach for explaining data. In: ICML (2021)
5. Cheng, C.-H., Fu, A., Zhang, F.: Entropy-based subspace clustering for mining numerical data. In: KDD (1999)
6. Cohen, S., Dror, G., Ruppin, E.: Feature selection via coalitional game theory. Neural Comput. **19**, 1939–1961 (2007)
7. Cover, T.M., Thomas, J.A.: Elements of Information Theory (Wiley Series in Telecommunications and Signal Processing). Wiley-Interscience, Hoboken (2006)
8. He, X., Cai, D., Niyogi, P.: Laplacian score for feature selection. In: NIPS (2006)
9. Lazo, A.V., Rathie, P.: On the entropy of continuous probability distributions. Trans. Inf. Theory **24**(1), 120–122 (1978)
10. Li, J., et al.: Feature selection: a data perspective. ACM Comput. Surv. **50**, 1–45 (2017)
11. Li, Z., Yang, Y., Liu, J., Zhou, X., Lu, H.: Unsupervised feature selection using nonnegative spectral analysis. In: AAAI (2012)
12. Lu, Y., Cohen, I., Zhou, X.S., Tian, Q.: Feature selection using principal feature analysis. In: MM (2007)
13. Lundberg, S.M., Lee, S.-I.: A unified approach to interpreting model predictions. In: NIPS (2017)
14. Nguyen, H., Müller, E., Vreeken, J., Keller, F., Böhm, K.: CMI: an information-theoretic contrast measure for enhancing subspace cluster and outlier detection. In: SDM (2013)
15. Pfannschmidt, K., Hüllermeier, E., Held, S., Neiger, R.: Evaluating tests in medical diagnosis: combining machine learning with game-theoretical concepts. In: Carvalho, J.P., Lesot, M.-J., Kaymak, U., Vieira, S., Bouchon-Meunier, B., Yager, R.R. (eds.) IPMU 2016. CCIS, vol. 610, pp. 450–461. Springer, Cham (2016). https://doi.org/10.1007/978-3-319-40596-4_38
16. Reshef, D., et al.: Detecting novel associations in large data sets. Science **334**, 1518–1524 (2011)
17. Rozemberczki, B., et al.: The Shapley value in machine learning (2022)
18. Shapley, L.S.: A value for N-person games. Contributions to the Theory of Games (1953)
19. Shekar, A.K., Bocklisch, T., Sánchez, P.I., Straehle, C.N., Müller, E.: Including multi-feature interactions and redundancy for feature ranking in mixed datasets. In: Ceci, M., Hollmén, J., Todorovski, L., Vens, C., Džeroski, S. (eds.) ECML PKDD 2017. LNCS (LNAI), vol. 10534, pp. 239–255. Springer, Cham (2017). https://doi.org/10.1007/978-3-319-71249-9_15
20. Solorio-Fernández, S., Carrasco-Ochoa, J.A., Martínez-Trinidad, J.F.: A review of unsupervised feature selection methods. Artif. Intell. Rev. **53**(2), 907–948 (2019). https://doi.org/10.1007/s10462-019-09682-y
21. Strumbelj, E., Kononenko, I.: An efficient explanation of individual classifications using game theory. J. Mach. Learn. Res. **11**, 1–18 (2010)
22. van Campen, T., Hamers, H., Husslage, B., Lindelauf, R.: A new approximation method for the Shapley value applied to the WTC 9/11 terrorist attack. Soc. Netw. Anal. Min. **8**, 1–12 (2018)

23. Vergara, J.R., Estévez, P.A.: A review of feature selection methods based on mutual information. Neural Comput. Appl. **24**(1), 175–186 (2013). https://doi.org/10.1007/s00521-013-1368-0

24. Wang, S., Tang, J., Liu, H.: Embedded unsupervised feature selection. In: AAAI (2015)

25. Yang, Y., Shen, H., Ma, Z., Huang, Z., Zhou, X.: $l_{2,1}$-norm regularized discriminative feature selection for unsupervised learning. In: IJCAI (2011)

26. Zhao, Z., Liu, H.: Spectral feature selection for supervised and unsupervised learning. In: ICML (2007)

27. Zhu, X., Wang, Y., Li, Y., Tan, Y., Wang, G., Song, Q.: A new unsupervised feature selection algorithm using similarity-based feature clustering. Comput. Intell. **35**, 2–22 (2019)

Multi-label Online Streaming Feature Selection Algorithms via Extending Alpha-Investing Strategy

Tianqi Ji, Xizhi Guo, Yunqian Li, Dan Li, Jun Li, and Jianhua Xu[✉]

School of Computer and Electronic Information, School of Artificial Intelligence,
Nanjing Normal University, Nanjing 210023, Jiangsu, China
{182202009,152202010,19190114,28190127}@stu.njnu.edu.cn,
{lijuncst,xujianhua}@njnu.edu.cn

Abstract. Multi-label learning is a special supervised pattern classification issue, in which an instance is possibly associated with multiple class labels simultaneously. As various real world applications emerge continuously in the big data field, more attention has been paid to streaming data forms recently, i.e., instance, feature and label streams. In this paper, we focus on multi-label online streaming feature selection (OSFS) problem, whose features arrive sequentially over time, and instances and labels are given, to choose an optimal subset of features dynamically. Alpha-investing method is one of the most cited embedded-type single-label OSFS techniques, which mainly involves the linear regression model as its classifier. In this paper, we generalize such a technique to build two new multi-label OSFS algorithms (simply ML-AIBR and ML-AIMOR), which are based on binary relevance (BR) decomposition way and multi-output regression (MOR), respectively. To the best of our knowledge, such two algorithms are the first proposed embedded-type OSFS technique for multi-label streaming features so far. Our extensive experiments conducted on six benchmark data sets demonstrate that our two proposed methods performs better than three existing algorithms. Specially, our ML-AIMOR could filter out more irrelevant and redundant features effectively.

Keywords: Multi-label learning · Feature selection · Streaming features · Online learning · Alpha investing

1 Introduction

Multi-label learning is a particular supervised classification task in which an instance could belong to multiple class labels at the same time and thus the classes are no longer exclusive to one another [4,14,23]. Nowadays, such a paradigm covers many real-world applications, for example, text categorization,

Supported by Natural Science Foundation of China (NSFC) under Grants 62076134 and 62173186.

music emotion classification, image annotation, bioinformatics and so on [4,14]. Traditional multi-label learning mainly deals with an off-line learning setting where all data are available in advance for learning. Therefore, there exist two possible limitations. On the one hand, those existing multi-label classification methods are impractical for large-scale applications in big data area since all data are needed to be stored in memory. On the other hand, it is non-trivial to adapt those off-line classifiers to sequential data [14,14,23]. In this case, online multi-label learning paradigm is proposed, which is to cope with various streaming data [28].

In multi-label learning, a collected data set is usually described via three dimensions: instance, feature and label, which correspondingly results in three streaming data forms: instance stream [19,28] (directly as data stream widely), label stream [10,11], and feature stream [8,9], when one dimension extends sequentially over time and the remained two dimensions are fixed.

As various real-world applications emerge continuously, their feature dimension goes much and much higher, which inevitably includes some irrelevant, redundant and noisy features. This situation usually costs more computational resources, and builds more complicated learning models and even deteriorates classification performance. To this end, lots of dimensionality reduction methods have been introduced under off-line setting in the past ten years, which involves feature extraction (FE) [21,24,25] and feature selection (FS) [6,18,21]. The former is to construct some secondary features via combining original features linearly or nonlinearly, and the latter to choose a most discriminative subset from all given original features. In order to remain feature physical meanings and enhance learning model interpretability, some researchers pay more attention to FS techniques and further extend them from static data to streaming ones [9,10]. In this paper, we focus on online streaming feature selection methods (OSFS), which tackles the feature selection problem under an online manner, where the features arrive sequentially over time.

In [9], according to fuzzy mutual information as relevance and redundancy evaluation indices, an OSFS method (i.e., MSFS) is proposed, which consists of two steps: online relevance analysis to decide whether a new arrived feature is selected or discarded, and online redundancy analysis to remove a redundant feature among selected features. Following this two-step analysis principle, the OM-NRS [12] substitutes neighborhood rough set for fuzzy mutual information [9]. The ML-OSMI is built via combining label spectral granulation with mutual information in [22]. In the I-SFS [17], via mutual information based relevance and redundancy, two objectives are optimized by multi-objective cuckoo search techniques, i.e., maximizing the difference between the relevance and redundancy in the selected features and minimizing difference between the redundancy to the relevance of unselected features.

The aforementioned three methods assume that the features arrive one-by-one over time. When a group of features are extracted sequentially, several group feature selection methods are built. With neighborhood symmetrical uncertainty and neighborhood mutual information, the literature [13] proposed a group OSFS method (OMGFS). Via integrating neighborhood mutual information, dynamic sliding window and feature repulsion loss, Li et al. are to construct another group OSFS approach (SF-DSW-FRL) [8]. In [16], a three-phase filtering process is proposed, where an evolutionary-based particle swarm optimization (PSO) is applied to the group of incoming features in a multi-objective optimization setting, the redundancy of features selected in the current group to the already selected features is checked, and those features in the already selected feature list that becomes non-significant on the selection of newly arrived features are discarded. In [17], the multi-objective cuckoo search technique is alternatively applied to the group of incoming features to build a corresponding G-SFS from I-SFS.

Usually, existing FS methods for both single-label OSFS [1,5] and off-line multi-label FS [6,18], are generally categorized into three groups: embedded, wapper and filter. According to this partitioning rule, the aforementioned multi-label OSFS methods all belong to filter-type FS techniques. Therefore, designing and implementing some non-filter multi-label OSFS are still an open issue.

Among existing single-label OSFS methods, alpha-investing technique [29] has become one of the most widely-recited embedded-type FS methods [1,5], which is based on linear regression, likelihood ratio test and adaptive threshold. In this paper, we generalize such an approach to deal with multi-label OSFS problem via two ways. One is to decompose a multi-label problem into several binary subproblems via binary relevance (BR) trick [4] at first, then execute single-label alpha-investing procedure for a new arrived feature independently for each binary subproblem, and finally decide whether to choose this feature. The other is to consider the multi-label problem as a multi-output regression (MOR) problem to replace single-output regression one in alpha-investing. Therefore, two multi-label OSFS methods are proposed in this paper, which are concisely referred to as ML-ALBR and ML-ALMOR respectively. To the best of our knowledge, such two multi-label OSFS algorithms are the first proposed embedded-type OSFS techniques so far. Our extensive experiments on six benchmark data sets demonstrate that our proposed OSFS methods are more effective, compared with three existing FS techniques (i.e., PMU [7], RF-ML [20] and MLNB [26]).

The rest of this paper is organized as follows. The Sect. 2 is to construct our new embedded-type multi-label OSFS methods (i.e., ML-AIBR and ML-ALMOR). The extensive experiments are conducted on six benchmark data sets in Sect. 3. Finally, our conclusions and future work are given in Sect. 4.

2   Multi-label Online Streaming Feature Selection Algorithms via Extending Alpha-Investing Strategy

This section consists of four sub-sections: preliminaries, single-label alpha-inverting method, and two multi-label OSFS methods respectively based on binary relevance decomposition and multi-output regression, to introduce two novel multi-label online streaming feature selection (OSFS) methods.

2.1 Preliminaries

On multi-label OSFS, we assume that the size of training set (N) and the number of class labels (C) are fixed, whereas the dimensionality of features (D) is increasing gradually over time and even goes to be infinite. Let the dynamic feature index set be $\mathcal{F} = \{f_1, ..., f_t, ...\}$ in which the feature f_t arrive at time t, and the static label index set be $\mathcal{L} = \{l_1, ..., l_C\}$. At time t, a multi-label training data can be depicted using a dynamic real feature matrix $\mathbf{X} \in \mathcal{R}^{N \times t}$

$$\mathbf{X}_t = [\mathbf{x}_{t1}, ..., \mathbf{x}_{ti}, ..., \mathbf{x}_{tN}]^T = [\mathbf{x}^1, ..., \mathbf{x}^j, ..., \mathbf{x}^t] \tag{1}$$

and a static binary label matrix $\mathbf{Y} \in \{0,1\}^{N \times C}$

$$\mathbf{Y} = [\mathbf{y}_1, ..., \mathbf{y}_i, ..., \mathbf{y}_N]^T = [\mathbf{y}^1, ..., \mathbf{y}^j, ..., \mathbf{y}^C] \tag{2}$$

where the i-th instance is represented by its column feature vector $\mathbf{x}_{ti} = [x_{t1}, ..., x_{ti}]^T \in \mathcal{R}^t$ and label vector $\mathbf{y}_i = [y_{i1}, ..., y_{iC}]^T \in \{0,1\}^C$, $\mathbf{x}^t \in R^N$ is the t-th arrived feature vector at time t, and $\mathbf{y}^j \in \{1,0\}^N$ is the j-th label vector.

Since the features flow continuously one-by-one, multi-label OSFS is to choose an optimal discriminative subset of the predefined size d from the available feature set at time t (generally $d < t$), via removing those irrelevant and redundant features, which finally is used to learn a multi-label classifier: $g(\mathbf{x}) : R^d \rightarrow \{0,1\}^C$, to predict the binary labels for unseen instances.

2.2 Alpha-Investing Method for Single-Label OSFS

Alpha-investing algorithm [29] is a typical single-label OSFS technique [1,5], which consists of three key factors: linear regression, likelihood ratio test and variable threshold (i.e., α).

For single-label classification problem, the above label matrix (2) is reduced into a column vector $\mathbf{y} = [y_1, ..., y_N]^T$, and at time t the linear regression [3]

$$y = \mathbf{w}_t^T \bar{\mathbf{x}}_t + e_t \tag{3}$$

is regarded as a linear classifier, where $\bar{\mathbf{x}}_t = [1, \mathbf{x}_t^T]^T$ is an extended instance feature vector, $\mathbf{w}_t \in R^{t+1}$ is weight vector including a bias term, and e_t is model error satisfying the standard normal distribution with the variance σ_t^2 (i.e., $N(0, \sigma_t^2)$).

Based on the feature matrix (1) and label vector $(\mathbf{y})$, a squared error sum is defined as

$$E_t = \left\| \mathbf{y} - \bar{\mathbf{X}}_t \mathbf{w}_t \right\|_2^2 \tag{4}$$

where $\| \cdot \|_2$ is 2-norm of vector, and $\bar{\mathbf{X}}_t = [\mathbf{1}, \mathbf{X}_t] \in R^{N \times (t+1)}$ and $\mathbf{1}$ is an unit column vector with all one elements.

Let the gradient vector of E_t with respect to $\mathbf{w}_t$ be zero (i.e., $\nabla E_t = \mathbf{0}$), the optimal weight vector $\mathbf{w}_t^*$ is obtained as follows

$$\mathbf{w}_t^* = \left[\bar{\mathbf{X}}_t^T \bar{\mathbf{X}}_t \right]^{-1} \bar{\mathbf{X}}_t^T \mathbf{y}. \tag{5}$$

Correspondingly, the minimum squared error sum becomes

$$E_t^* = \left\| \mathbf{y} - \tilde{\mathbf{X}}_t \mathbf{w}_t^* \right\|_2^2 \qquad (6)$$

and the variance is

$$\sigma_t^2 = \frac{E_t^*}{N} \qquad (7)$$

Assume that one feature subset $\mathcal{S}_t$ is chosen at time t. When at time $t+1$, a new feature f_{t+1} arrives, its corresponding E_{t+1}^* is estimated via (6) similarly. In this case, the probability p-value of likelihood ration test is calculated as

$$p_{t+1} = \frac{\left(\frac{1}{\sqrt{2\pi\sigma_t^2}} \right)^t \exp\left(-\frac{E_t^*}{2\sigma_t^2} \right)}{\left(\frac{1}{\sqrt{2\pi\sigma_t^2}} \right)^t \exp\left(-\frac{E_{t+1}^*}{2\sigma_t^2} \right)} = \exp\left(\frac{E_{t+1}^* - E_t^*}{2\sigma_t^2} \right) \qquad (8)$$

which implies that a smaller p-value will indicate a greater contribution of this arrived feature to the linear regression model. When this p-value is less than a threshold α (i.e., $p_{t+1} < \alpha$), this arrived feature would be selected, i.e., $\mathcal{S}_{t+1} = \mathcal{S}_t \cup \{f_{t+1}\}$.

Further, alpha-investing algorithm adjusts this threshold α dynamically. Let β_0 be an initial probability of false positive. At time $t+1$, the threshold is set to $\alpha_t = \beta_t/2(t+1)$ at first. If the $p_{t+1} < \alpha_t$ and the feature f_{t+1} is chosen, the β is increased as follows

$$\beta_{t+1} = \beta_t + \alpha_\delta - \alpha_t \qquad (9)$$

where $\alpha_\delta < 1$ is a user predefined constant, otherwise

$$\beta_{t+1} = \beta_t - \alpha_t, \qquad (10)$$

which decreases the threshold when this feature is discarded. Such a alpha-investing rule is to control the threshold (α) adaptively for adding new arrived features, so that when a new feature is added to regression model, one invests α increasing the wealth, raising the threshold, and allowing a slightly higher future chance of incorrect inclusion of sequential features.

Therefore alpha-investing algorithm combines linear regression, likelihood ratio test and floating threshold to build an embedded-type OSFS approach. In addition, two user-defined constants are set to $\beta_0 = 0.5$ and $\alpha_\delta = 0.5$ in its experiments [29]. In the next two sub-sections, we will generalize this alpha-investing method to tackle multi-label OSFS issue.

2.3   Multi-label Online Streaming Feature Selection via Combining Alpha-Investing with Binary Relevance

In multi-label learning, a widely-used strategy is to partition a multi-label classification problem into one or more single-label classification subproblems that are dealt with using various single-label techniques [4,14,23]. Among existing

Algorithm 1. Multi-label online streaming feature selection with alpha-investing and binary relevance (ML-AIBR)

Input: Label matrix Y, and initial false probabilities $\beta_0 = 0.5$, and $\alpha_\delta = 0.5$
Procedure
Initialize the feature matrix as an unit column vector $\mathbf{X}_0 = [\mathbf{1}]$,
    selected feature subset $\mathcal{S}_0 = \{\}$, $\beta_0^j = \beta_0 (j = 1, 2, ..., C)$ and $t = 0$
Do the following loop procedure
    Get a new feature f_{t+1} with its feature vector $\mathbf{x}^t$ at time $t + 1$
    Update the feature matrix $\mathbf{X}_{t+1} = [\mathbf{X}_t, \mathbf{x}^t]$
    For $j = 1, 2, ..., C$
        Construct a training subset $\{\mathbf{X}_{t+1}, \mathbf{y}^j\}$ for the k-the label
        Set $\alpha_t^j = \beta_t^j / 2(t + 1)$.
        Calculate its p-value p_{t+1}^j
        If $p_{t+1}^j < \alpha_t^j$ (to choose this feature)
            Add this feature subset, $\mathcal{S}_{t+1} = \mathcal{S}_t \cup \{f_{t+1}\}$
            Update $\beta_{t+1}^j = \beta_t^j + \alpha_\delta - \alpha_t^j$
        Else (to discard this feature)
            Update $\beta_{t+1}^j = \beta_t^j - \alpha_t^j$
        End if
    End for
    $t = t + 1$
Until no feature arrives
Output: Selected feature subset $\mathcal{S}$

---

decomposition ways, binary relevance (BR) is the most popular decomposition strategy, which separates a C-label problem into C independent binary subproblems. For the above training set ((1) and (2)), at time t, we could obtain C binary subsets:

$$\{\mathbf{X}_t, \mathbf{y}^j | j = 1, 2, .., C\}. \tag{11}$$

We execute the feature selection step in alpha-investing method for each binary subset, and then decide whether such an arrived feature f_{t+1} is selected or not. Once this feature f_{t+1} is selected from the j-th binary subset, we add such a feature to selected feature subset, i.e., $\mathcal{S}_{t+1} = \mathcal{S}_t \cup \{f_{t+1}\}$ and increase its corresponding threshold β_t^j according to (9). Otherwise, only the threshold β_t^j is decreased adaptively via (10). Finally, we construct an embedded-type multi-label OSFS method via combining alpha-investing and binary relevance, which is simplified as ML-AIBR in this paper and summarized in Algorithm 1.

2.4 Multi-label Online Streaming Feature Selection via Combining Alpha-Investing with Multi-output Regression

Multi-output regression [2] is a generalized form of the classical single-output regression, which is to deal with more than one real output problem. Generally, multi-label classification with multiple binary labels could also be regarded as a special case of multi-output regression [15].

Algorithm 2. Multi-label online streaming feature selection with alpha-investing and multi-output regression (**ML-AIMOR**)

Input: Label matrix $\mathbf{Y}$, and initial false probabilities $\beta_0 = 0.5$ and $\alpha_\delta = 0.5$
Procedure
Initialize the feature matrix $\mathbf{X}_0 = [\mathbf{1}]$, selected feature subset $\mathcal{S}_0 = \{\}$ and $t = 0$
Do the following loop procedure
    Get a new feature f_{t+1} with its feature vector $\mathbf{x}^t$ at time $t + 1$
    Update the feature matrix $\mathbf{X}_{t+1} = [\mathbf{X}_t, \mathbf{x}^t]$
    Construct a training set $\{\mathbf{X}_{t+1}, \mathbf{Y}\}$
    Set $\alpha_t = \beta_t / 2(t + 1)$.
    Calculate its p-value p_{t+1} using (18)
    If $p_{t+1} < \alpha_t$ (to choose this feature)
        Add this feature subset, $\mathcal{S}_{t+1} = \mathcal{S}_t \cup \{f_{t+1}\}$
        Adjust $\beta_{t+1} = \beta_t + \alpha_\delta - \alpha_t$
    Else (to discard this feature)
        Adjust $\beta_{t+1} = \beta_t - \alpha_t$
    End if
    $t = t + 1$
Until no feature arrives
Output: Selected feature subset $\mathcal{S}$

In this sub-section, we will substitute multi-output regression for singe-output one to extend alpha-investing for multi-label OSFS case.

For the above multi-label training set ((1) and (2)), we define C regression functions:

$$y^j = \mathbf{w}_{jt}^T \bar{\mathbf{x}}_t + e_{jt}, j = 1, ..., C \tag{12}$$

as a multi-label linear classifier, where $\mathbf{w}_{jt} \in R^{t+1}$ is a weight vector for the j-th label, and e_{jt} is the j-th model error satisfying the standard normal distribution with the variance σ_{jt}^2 (i.e., $N(0, \sigma_{jt}^2)$).

Correspondingly, the total squared error sum is depicted as follows

$$E_t = \frac{1}{C} \sum_{j=1}^{C} \left\| \mathbf{y}^j - \bar{\mathbf{X}}_t \mathbf{w}_{jt} \right\|_2^2 = \frac{1}{C} \left\| \mathbf{Y} - \bar{\mathbf{X}}_t \mathbf{W}_t \right\|_F^2 \tag{13}$$

with

$$\mathbf{W}_t = [\mathbf{w}_{1t}, \mathbf{w}_{2t}, ..., \mathbf{w}_{Ct}] \in R^{(t+1) \times C} \tag{14}$$

where $\| \cdot \|_F$ is the Frobenius norm of matrix. Similarly, the optimal weight matrix $\mathbf{W}_t^*$ is achieved

$$\mathbf{W}_t^* = \left[\bar{\mathbf{X}}_t^T \bar{\mathbf{X}}_t \right]^{-1} \bar{\mathbf{X}}_t^T \mathbf{Y} \tag{15}$$

and then the minimum total squared error sum and the total variance are estimated as

$$E_t^* = \frac{1}{C} \left\| \mathbf{Y} - \bar{\mathbf{X}}_t \mathbf{W}_t^* \right\|_F^2 \tag{16}$$

and

$$\sigma_t^2 = \frac{E_t^*}{N} \tag{17}$$

Table 1. Statistical information of six used multi-label data sets.

Data set	Training instances	Testing instances	Features	Labels	Label cardinality
Arts	2000	3000	462	26	1.636
Business	2000	3000	438	30	1.588
Health	2000	3000	612	32	1.662
Recreation	2000	3000	606	22	1.423
Reference	2000	3000	793	33	1.169
Social	2000	3000	1047	39	1.283

At time $t+1$, a new feature f_{t+1} arrives, we utilize E_t^* (16) and σ_t^2 (17) to replace these two quantities in (6) and (7) to calculate the p-value

$$p_{t+1} = \exp\left(\frac{E_{t+1}^* - E_t^*}{2\sigma_t^2}\right) \tag{18}$$

to decide whether to select this feature f_{t+1}, and then to adjust the β value adaptively according to (9) and (10). We integrate the single-label alpha-investing method with the multi-label regression to build a new multi-label OSFS algorithm, which is simply named as ML-ALMOR, and is summarized in Algorithm 2. Compared with the aforementioned ML-AIBR, this ML-AIMOR considers multi-label OSFS problem as an entire ones, rather than several separated subproblems.

3   Experiments

In this section, we will evaluate our proposed two multi-label OSFS algorithms (ML-AIBR and ML-ALMOR), and three existing FS methods (PMU [7], RF-ML [20] and MLNB [26], on six multi-label benchmark data sets.

In our experiments, we download six Yahoo benchmark data sets: Arts, Business, Health, Recreation, Reference and Social, from Mulan library[1]. The Table 1 lists their main statistical information, including the numbers of training and testing instances, dimensionality of features, number of labels, and label cardinality (i.e., average label size). These data sets have been widely used to evaluate multi-label OSFS techniques in [9,12,13]. Although these data sets are originally not streaming feature format, we regard them as streaming features over time via their natural arranged orders, as in [9,12,13], to conduct our comparison experiments.

On the other hand, we choose three FS methods (PMU [7], ReliefF-ML [20] and MLNB [26] as our compared techniques, which have been used in the experiments comparison in [9,12,13,16,17]. PMU [7] implements a FS procedure by

---

[1] http://mulan.sourceforge.net/datasets-mlc.html.

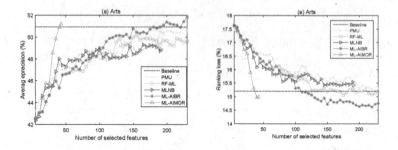

Fig. 1. Two metrics with the different number of selection features on arts

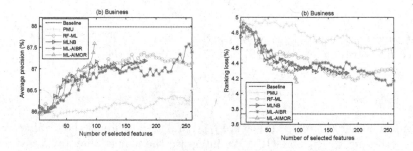

Fig. 2. Two metrics with the different number of selection features on business

maximizing the dependency between features and labels measured by multivari-ate mutual information. RF-ML [20] extends traditional single-label FS approach ReliefF to deal with multi-label FS problem, whose smoothing factor is set to 1 and the number of nearest neighbor instances is 10. MLNB [26] uses PCA to eliminate irrelevant or redundant features via some smallest eigenvalues, then select an optimal feature subset via genetic algorithm.

After executing FS procedure, multi-label k-nearest neighbor classifier (ML-kNN) [27] is used as our baseline classifier, whose smoothing factor is 1 and $k = 10$. There are more than 20 classification performance evaluation metrics in [4]. Due to the limited space, in this study, we choose two metrics (average precision and ranking loss) to compare and evaluate our two proposed methods (ML-ALBR and ML-ALMOR), and three existing techniques (PMU, RF-ML and MLNB). For a well-performed FS technique, its corresponding average precision is high, whereas the ranking loss is low.

We investigate these two metrics as functions of the different number of selected features, where the numbers of selected features are from 5 to the fixed number (230, 260, 330, 301, 410, 460 for six data sets on Table 1 according to their different size of original feature sets) with a step 5.

The experiments are shown in Figs. 1, 2, 3, 4, 5 and 6 where the left and right subplots corresponds to average precision and ranking loss, respectively. From these figures, we observe that:

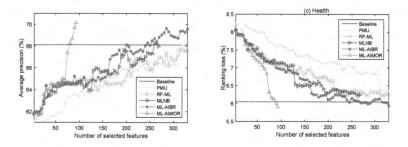

Fig. 3. Two metrics with the different number of selection features on health

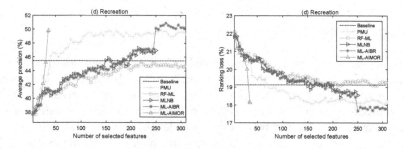

Fig. 4. Two metrics with the different number of selection features on recreation

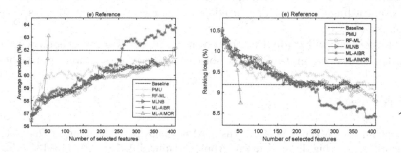

Fig. 5. Two metrics with the different number of selection features on reference

(i) As the number of selected features increases, the average precision values increase and the ranking loss decrease gradually, respectively.

(ii) In order to outperform three existing feature selection methods, our proposed method ML-ALBR selects more features.

(iii) Compared with three existing methods and ML-AIBR, our proposed algorithm ML-AIMOR only selects fewer features, to achieve a competitive performance.

These experimental results demonstrate that our two proposed online streaming feature selection methods are effective, compared with three existing feature selection techniques.

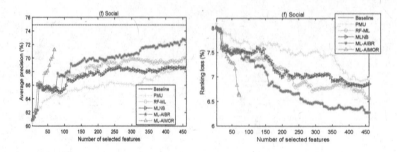

Fig. 6. Two metrics with the different number of selection features on social

4   Conclusions

In this paper, we extend the traditional online streaming feature selection method: alpha-investing, through two effective ways: binary relevance decomposition and multi-output regression to deal with multi-label streaming feature selection problems, resulting in two effective multi-label online streaming feature selection approaches. To the best of our knowledge, they are two first proposed embedded-type online feature selection algorithms for streaming features. Our extensive experiments on six benchmark data sets validate the effectiveness of our proposed techniques, compared with three existing feature selection techniques, according to two widely-used performance evaluation metrics (average precision and ranking loss).

In future, we will evaluate and compare our proposed methods with more benchmark data sets and more state-of-the-art feature selection approaches.

References

1. Alnuaimi, N., Masud, M.M., Serhsni, M.S., Zaki, N.: Streaming feature selection algorithms for big data: a survey. Appl. Comput. Inf. **18**(1/2), 113–135 (2022)
2. Borchani, H., Varando, G., Bielza, C.: A survey on multi-output regression. Wires. Data Min. Knowl. **5**(5), 216–233 (2015)
3. Freeedman, D.A.: Statistical Models: Theory and Practice. Cambridge University Press, Cambridge (2009)
4. Herrera, F., Charte, F., Rivera, A.J., del Jesus, M.J.: Multilabel Classification Problem Analysis, Metrics and Techniques. Springer, Cham (2016). https://doi.org/10.1007/978-3-319-41111-8
5. Hu, X., Zhou, P., Li, P., Wang, J., Wu, X.: A survey on online feature selection with streaming features. Front. Comput. Sci. **12**(3), 479–493 (2018). https://doi.org/10.1007/s11704-016-5489-3
6. Kashef, S., Nezamabadi-pour, H., Nipour, B.: Multilabel feature selection: a comprehensive review and guide experiments. WIREs Data Min. Knowl. Discovery **8**(2), Article ID e1240 (2018)

7. Lee, J., Kim, D.W.: Feature selection for multi-label classification using multivariate mutual information. Pattern Recogn. Lett. **34**(3), 349–357 (2013)
8. Li, Y., Cheng, Y.: Streaming feature selection for multi-label data with dynamic sliding windows and feature repulsion loss. Entropy **21**, Article ID 1151 (2019)
9. Lin, Y., Hu, Q., Liu, J., Wu, X.: Streaming feature selection for multi-label learning based on fuzzy mutual information. IEEE Trans. Fuzzy Syst. **25**(6), 1491–1507 (2017)
10. Lin, Y., Hu, Q., Zhang, J., Wu, X.: Multi-label feature selection with streaming labels. Inf. Sci. **372**, 256–275 (2016)
11. Liu, J., Li, Y., Weng, W., Zhang, J., Chen, B., Wu, S.: Feature selection for multi-label learning with stream label. Neurocomputing **387**, 268–278 (2020)
12. Liu, J., Lin, Y., Li, Y., Weng, W., Wu, S.: Online multi-label streaming feature selection based on neighborhood rough set. Pattern Recogn. **84**, 273–287 (2018)
13. Liu, J., Lin, Y., Wu, S., Wang, C.: Online multi-label group feature selection. Knowledge-Based Syst. **143**, 42–57 (2018)
14. Liu, W., Shen, X., Wang, H., Tsang, I.W.: The emerging trends of multi-label learning. IEEE Trans. Pattern Anal. Mach. Intell. (2021). https://doi.org/10.1109/TPAMI.2021.3119334
15. Osojnik, A., Panov, P., Džeroski, S.: Multi-label classification via multi-target regression on data streams. Mach. Learn. **106**(6), 745–770 (2016). https://doi.org/10.1007/s10994-016-5613-5
16. Paul, D., Jain, A., Saha, S., Mathew, J.: Multi-objective PSO based online feature selection for multi-label classification. Knowl.-Based Syst. **222**, e106966 (2021)
17. Paul, D., Kumar, R., Saha, S., Mathew, J.: Multi-objective cuckoo search-based streaming feature selection for multi-label dataset. ACM Trans. Knowl. Discov. Data **15**(6), e93 (2021)
18. Pereira, R.B., Plastino, A., Zadrozny, B., Merschmann, L.H.C.: Categorizing feature selection methods for multi-label classification. Artif. Intell. Rev. **49**(1), 57–78 (2016). https://doi.org/10.1007/s10462-016-9516-4
19. Read, J., Bifet, A., Holmes, G., Pfahringer, B.: Scalable and efficient multi-label classification for evolving data streams. Mach. Learn. **88**(1/2), 243–272 (2012)
20. Reyes, O., Morell, C., Ventura, S.: Scalable extensions of the ReliefF algorithm for weighting and selecting features on the multi-label learning context. Neurocomputing **161**, 168–182 (2015)
21. Siblini, W., Kuntz, P., Meyer, F.: A review on dimensionality reduction for multi-label classification. IEEE Trans. Knowl. Data Eng. **33**(3), 839–857 (2021)
22. Wang, H., Yu, D., Li, Y., Li, Z., Wang, G.: Multi-label online streaming feature selection based on spectral granulation and mutual information. In: Nguyen, H.S., Ha, Q.-T., Li, T., Przybyła-Kasperek, M. (eds.) IJCRS 2018. LNCS (LNAI), vol. 11103, pp. 215–228. Springer, Cham (2018). https://doi.org/10.1007/978-3-319-99368-3_17
23. Wever, M., Tornede, A., Mohrand, F., Hullermeier, E.: AutoML for multi-label classification: overview and empirical evaluation. IEEE Trans. Pattern Anal. Mach. Intell. **43**(9), 3037–3054 (2021)
24. Xu, J.: A weighted linear discriminant analysis framework for multi-label feature extraction. Neurocomputing **275**, 107–120 (2018)
25. Xu, J., Mao, Z.H.: Multilabel feature extraction algorithm via maximizing approximated and symmetrized normalized cross-covariance operator. IEEE Trans. Cybern. **51**(7), 3510–3523 (2021)
26. Zhang, M.L., Peña, J.M., Robles, V.: Feature selection for multi-label Naive Bayes classification. Inf. Sci. **179**(19), 3218–3229 (2009)

27. Zhang, M.L., Zhou, Z.H.: ML-kNN: a lazy learning approach to multi-label learning. Pattern Recogn. **40**(7), 2038–2048 (2007)
28. Zheng, X., Li, P., Chu, Z., Hu, X.: A survey on multi-label data stream classification. IEEE Access **8**, 1249–1275 (2020)
29. Zhou, J., Foster, D., Stine, R., Ungar, L.: Streaming feature selection using alpha-investing. In: 7th ACM SIGKDD International Conference on Knowledge Discovery in Data Mining (KDD 2005), pp. 384–393. ACM Press, New York (2005)

Feature Selection Under Fairness and Performance Constraints

Ginel Dorleon[⊠][iD], Imen Megdiche[iD], Nathalie Bricon-Souf[iD],
and Olivier Teste[iD]

Toulouse Institute for Computer Science Research (IRIT), Toulouse, France
{ginel.dorleon,imen.megdiche,nathalie.bricon-souf,olivier.teste}@irit.fr

Abstract. Feature selection is an essential preprocessing procedure in data analysis. The process refers to selecting a subset of relevant features to improve prediction performance and better understand the data. However, we notice that traditional feature selection methods have limited ability to deal with data distribution over protected features due to data imbalance and indeed protected features are selected. Two problems can occur with current feature selection methods when protected features are considered: the presence of protected features among the selected ones which often lead to unfair results and the presence of redundant features which carry potentially the same information with the protected ones. To address these issues, we introduce in this paper a fair feature selection method that takes into account the existence of protected features and their redundant. Our new method finds a set of relevant features with no protected features and with the least possible redundancy under prediction quality constraint. This constraint consists of a trade-off between fairness and prediction performance. Our experiments on well-known biased datasets from the literature demonstrated that our proposed method outperformed the traditional feature selection methods under comparison in terms of performance and fairness.

Keywords: Feature selection · Fairness · Protected features · Bias · Machine learning

1 Introduction

Feature selection (FS) is a popular dimensionality reduction technique for processing large dataset. The main objective of any FS method is to select a subset of relevant features from the input data that helps improving model's prediction. Fairness is another quality of the prediction model which can be of high importance for the usability of the model. Some specific features, known as protected, could induce problems when dealing with fairness and it has been proved [1] that protected features can lead to unfair decisions against minority groups.

According to [2], protected features are features that are of particular importance either for social, ethical or legal reasons when making decisions. Some examples of protected features are: sex, race, age, religion. With existing feature

R. Wrembel et al. (Eds.): DaWaK 2022, LNCS 13428, pp. 125–130, 2022.
https://doi.org/10.1007/978-3-031-12670-3_11

selection methods, two major problems are identified among the selected features which are: (1) protected features whose presence leads to biased results and (2) the presence of redundant features to the protected whose deletion leads to a loss in the prediction performance. In this study, redundancy is considered in the sense of correlation between non-independent features and the fact that the latter can be strongly correlated with others enabling a classifier to reconstruct them. Thus, in our work, we focused on these two problems identified among selected features that directly affect performance and fairness. In order to solve this, we introduce a method that allows to obtain the best trade-off between performance and fairness. Our method finds a set of relevant features without protected feature and with the least possible redundancy which maximizes the performance while ensuring fairness of the model obtained.

For our contributions in this work: (1) we introduce a more flexible way to use threshold for redundancy analysis by defining a threshold space instead of using a single value which could be subjective, (ii) we define an outcome-fairness algorithm for dealing with protected features in decision support algorithm which takes into consideration redundant features while making decisions on fairness, so that the overall performance remains high.

In Sect. 2, we summarize the different existing methods to tackle the issues identified with their limitations. Section 3 presents our new approach to deal with protected features, redundancy and fairness issues. The experimental results are described and analyzed in Sect. 4 of the full paper available here.

2   Related Work

Various feature selection methods were proposed to deal with the problem of redundancy and protected features. Here we look at those who deal with redundancy analysis and those who deal with protected features.

For the first category, authors have introduced in [3–5] different strategies to deal with redundancy in feature selection. However, when analyzing the methods cited above, we found that they inappropriately remove redundancy because they require users to set a single-defined threshold. As so, feature redundancy depends on the threshold set, that being said, different thresholds led to different sets of redundant features; thus, different models.

For the second category, we noticed the work in [6,7] where different strategies were used to handle the problem posed by protected features. These methods mostly focus on removing completely all the protected features. Again, we noticed various limitations to the approaches cited above trying to improve fairness while considering protected features. The approach of [6] of completely removing protected features may not solve the problem because there may be redundant features or even proxies to the protected that can reveal the same information. We notice the same observation for the work in [7], where redundant features to the protected are also ignored; this is dangerous in terms of fair outcomes when dealing with decisions problems involving minority groups.

Given the limitations of these above methods, there is a need for more in-depth research to overcome these limitations. Thus, we propose a new feature selection method which allows the building of efficient and fair models without protected feature and with the least possible redundancy. Our new method is a trade-off between performance and fairness.

3   The Proposed Method

Here we present our approach with the different steps are illustrated in Fig. 1. Our method takes as input a dataset divided into protected and unprotected features. Then, it performs a redundancy analysis based on a defined threshold space (S). Following the redundancy analysis, two subsets of features are obtained: a list of non redundant features (N) and a list of redundant features (R). These two lists are used subsequently to train various models using all possible partitions between (N) and (R). The partitions are created by adding combinations without duplication from (R) to (N). Each partition is used to train a model, then for each model obtained, we calculate its f-score, its fairness and a trade-off score (Δ). We will keep as final model the one which has the highest trade-off (delta) score, i-e, the most efficient and fair one. With this new method, we propose an efficient solution to the problem related to protected and redundant features on performance and fairness. This method makes it possible to take into account i) redundancy, ii) protected features and iii) fairness. For a detailed explanation of the proposed method, please read the full paper here.

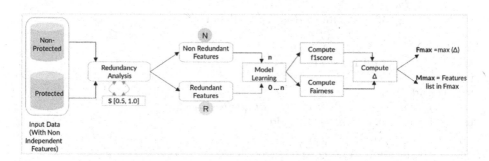

Fig. 1. The proposed approach and its different stages

3.1   Algorithm of the Proposed Method

Algorithm 1 shows the process of the proposed method. Let N be the set of non-redundant features, we denote by f_i a feature of N. Let P the set of known protected features or designated by a system expert before any analysis, we denote by p_i any feature of P. The algorithm takes as input two lists: the list of non-protected features N and the list of protected features annotated P from the input dataset. We use the following defined parameters: t the step ($t = 0.05$) to iterate over S and the hyper-parameter space ($S = [0.5, 1.0]$). We start by initializing the values of $Fmax$, $Mmax$, d, t and the empty list R (line 2–6). For each hyper-parameter (threshold) d in S, we compute redundancy between the list P and N, if any redundant feature is found according to d, it is added to R then removed from N (line 7–16). Using N, we iteratively increment over all possible partitions cr of R to train all possible models using all the partitions between the lists N and R, evaluate them and calculate their delta according to the specified formula (17–25). Then we decrement, start over using a new value of d until all possible value in the hyper-parameter S have been used (line 27). For the output of the algorithm, we have **Fmax**, the max of all the delta, and **Mmax**: the list of features constituting the model which led to Fmax.

Algorithm 1: Pseudo-code of the proposed method

Input: N, P // Non protected and protected Features
Output: Fmax, Mmax //max performance & feature list
1 **Begin**
2   $t \leftarrow 0.05$ //*iteration step over* S
3   $d \leftarrow 1.0$ //*highest threshold value in* S
4   $R \leftarrow \{\,\}$ //*redundant list*
5   $Fmax = 0$ //*max(Δ) to maximize*
6   $Mmax \leftarrow \{\,\}$ //*feature list of Fmax*
7 **while** $d \in$ S **do**
8   //*finding redundant features*
9   **for** $f_i \in N$ **do**
10   **for** $p_i \in P$ **do**
11   **if** $|$compute corr$(p_i, f_i)| \geq$ d **then**
12                  $R \leftarrow \{f_i\} \cup R$
13                  $N \leftarrow N \setminus \{f_i\}$
14   **end if**
15   **end do**
16   **end do**
17   //*search for the best model with the best tradeoff*
18   **for** $cr \in$ partition(R) $\cup \{\,\}$ **do**
19   compute $\hat{f}$ using $N \cup cr$
20   compute Δ using eq. (1)
21   **if** $\Delta \geq$ Fmax **then**
22   Fmax $\leftarrow \Delta$
23   Mmax $\leftarrow \hat{f}$
24   **end if**
25   **end do**
26      $d \leftarrow$ d - t
27 **end do**
28 **End**

---

4   Experimental Setup

We carried out our experiments with the goal of comparing the results obtained with our method with other Feature Selection methods. For that, two other existing feature selection methods were used for comparison: mRMR [3] and FCBF [4]. In particular, this comparison was made based on 3 criterion (performance, fairness and the delta score) using a classification task.

To evaluate and compare the proposed method to existing methods, we proceeded to a learning task by considering a binary classification problem over the 4 datasets that we describe on Table 1. For this binary classification, Random-Forest [8] and AdaBoost [9] were used as classifiers. Each model is trained and evaluated using the classic cross-validation procedure. F1-score is used as measure to assess the performance of each trained model. For more details please read the full paper here.

Table 1. Experimental datasets used

Dataset	Observations	Features	Protected used
German credit scoring	100	9	1
Adult income	32561	15	2
Bank churn	10147	13	1
Loan approval	615	14	2

4.1   Results Analysis

The analysis of the results is based on two criterion: the number of selected features and the trade-off score (fairness and performance). In general, on these 4 datasets, we get satisfactory results and we have maintained a good level of performance (f1-score), a higher fairness guarantying a higher score for the trade-off between f1-score and fairness.

Overall, the results of the experiments show that our method performs well and our redundancy analysis guided by protected features gives a better performance value in term of fairness and the trade-off score generally. For a detailed results on selected features, read the full paper here.

5   Conclusion

In this article, we present a novel feature selection method to improve performance and fairness in the case of protected features while considering their redundant. To achieve our goal, we introduce a trade-off strategy between performance and fairness. This new method, unlike existing methods allows in the presence of protected and redundant features to obtain a model that is both optimal and fair. The performance of our method is experimentally evaluated on four well known biased datasets. Compared to two other existing feature selection methods, we obtain satisfactory results. The comparative results obtained show our method's effectiveness in boosting fairness while maintaining a high level of performance. Furthermore, with our method, we show that it is possible to comply with data privacy policy by not using protected features while remaining efficient and fair.

Our future work should focus data distribution over protected and redundant features and sort out the imbalance that can lead to bias.

Source Code: The full source code including data of our experiments is available on GitHub under request. This version of the work is a short paper, please read the full paper here for more details on the proposed method and results.

References

1. Yeom, S., Datta, A., Fredrikson, M.: Hunting for discriminatory proxies in linear regression models. In: Proceedings of the 32nd International Conference on Neural Information Processing Systems (NIPS 2018), pp. 4573–4583. Curran Associates Inc., Red Hook (2018)
2. Fang, B., Jiang, M., Cheng, P., Shen, J., Fang, Y.: Achieving outcome fairness in machine learning models for social decision problems. In: Bessiere, C. (ed.) Proceedings of the Twenty-Ninth International Joint Conference on Artificial Intelligence, IJCAI 2020, pp. 444–450 (2020). ijcai.org
3. Peng, H., Long, F., Ding, C.: Feature selection based on mutual information criteria of max-dependency, max-relevance, and min-redundancy. IEEE Trans. Pattern Anal. Mach. Intell. **27**, 8 (2005)
4. Lei, Yu., Liu, H.: Efficient feature selection via analysis of relevance and redundancy. J. Mach. Learn. Res. **5**(2004), 1205–1224 (2004)
5. Wang, M., Tao, X., Han, F.: A new method for redundancy analysis in feature selection. In: 2020 3rd International Conference on Algorithms, Computing and Artificial Intelligence (ACAI 2020), Article 21, 5 p. Association for Computing Machinery, New York (2020)
6. Dwork, C., Hardt, M., Pitassi, T., Reingold, O., Zemel, R.: Fairness through awareness. In: Proceedings of the 3rd Innovations in Theoretical Computer Science Conference (ITCS 2012), pp. 214–226. Association for Computing Machinery, New York (2012)
7. Yan, S., Kao, H., Ferrara, E.: Fair class balancing: enhancing model fairness without observing sensitive attributes. In: Proceedings of the 29th ACM International Conference on Information & Knowledge Management (CIKM 2020). Association for Computing Machinery, New York (2020)
8. Pal, M.: Random forest classifier for remote sensing classification. Int. J. Remote Sens. **26**(1), 217–222 (2005)
9. Schapire, R.E.: Explaining adaBoost. In: Schölkopf, B., Luo, Z., Vovk, V. (eds.) Empirical Inference, pp. 37–52. Springer, Heidelberg (2013). https://doi.org/10.1007/978-3-642-41136-6_5

Time Series Processing

Interpretable Input-Output Hidden Markov Model-Based Deep Reinforcement Learning for the Predictive Maintenance of Turbofan Engines

Ammar N. Abbas[1]([✉]) [iD], Georgios C. Chasparis[1] [iD], and John D. Kelleher[2] [iD]

[1] Software Competence Center Hagenberg, Hagenberg, Austria
{ammar.abbas,georgios.chasparis}@scch.at
[2] ADAPT Research Centre, Technological University of Dublin, Dublin, Ireland
john.d.kelleher@tudublin.ie

Abstract. An open research question in deep reinforcement learning is how to focus the policy learning of key decisions within a sparse domain. This paper emphasizes on combining the advantages of input-output hidden Markov models and reinforcement learning. We propose a novel hierarchical modeling methodology that, at a high level, detects and interprets the root cause of a failure as well as the health degradation of the turbofan engine, while at a low level, provides the optimal replacement policy. This approach outperforms baseline deep reinforcement learning (DRL) models and has performance comparable to that of a state-of-the-art reinforcement learning system while being more interpretable.

Keywords: Deep Reinforcement Learning (DRL) · Input-Output Hidden Markov Model (IOHMM) · Predictive maintenance · Interpretable AI

1 Introduction

Predictive maintenance can be categorized as (i) *Prognosis*: predicting failure and notifying for replacement or repair ahead of time (*Remaining Useful Life* or briefly RUL is usually used as a prognosis approach, which is the estimation of the remaining life of equipment or a system until it becomes non-functional [20]); (ii) *Diagnosis*: predicting the actual cause of failure in the future through cause-effect analysis, or (iii) *Proactive Maintenance*: anticipate and mitigate the failure modes and conditions before they develop [6]. While proactive maintenance captures the root cause of potential failure, predictive maintenance performs an

supported by Collaborative Intelligence for Safety-Critical systems (CISC) project; funded by the European Union's Horizon 2020 Research and Innovation Programme under the Marie Skłodowska-Curie grant agreement no. 955901. The work of Kelleher is also partly funded by the ADAPT Centre which is funded under the Science Foundation Ireland (SFI) Research Centres Programme (Grant No. 13/RC/2106_P2).

R. Wrembel et al. (Eds.): DaWaK 2022, LNCS 13428, pp. 133–148, 2022.
https://doi.org/10.1007/978-3-031-12670-3_12

overall data analytics to be able to ensure scheduled maintenance. In this paper, the aforementioned questions will be investigated in the context of predictive maintenance of turbofan engines [4,18].

Reinforcement Learning (RL) is a natural approach to solving time series-based stochastic decision problems, such as predictive maintenance [21], and has recently shown promising results. RL systems learn by interacting with the environment and can learn in an online setting without having the data set beforehand [22]. However, when the key policy decision learned by an RL agent is relatively rare in a data set (such as the decision of when to change the equipment before failure while maximizing its use), the policy can be dominated by irrelevant phenomena, resulting in inefficient training. At the same time, the derived optimal policy does not provide interpretations or the root cause of the failure, and therefore keeps humans out of the loop with limited collaborative intelligence. Furthermore, in real-world industrial environments, RL learns directly from the observed raw sensor data that does not provide information about the unobserved hidden factors responsible for the decision-making of the system such as its health, which can limit the RL agent to learning an optimal policy.

An Input-Output Hidden Markov Model (IOHMM) [2,17] is a form of Bayesian Network that involves probabilistic inference of latent variables. An IOHMM extends the standard HMM model by integrating the dependencies of various covariates (inputs) to the initial, transition, and emission probabilities [1]. It can overcome the challenges faced by RL through (i) learning unobserved states and interpretations based on those hidden states, (ii) combining multiple correlated sensor data, (iii) defining the state of the system and its hierarchical distribution based on its different levels of operation (normal, starting point of failure, close to failure, etc.), and (iv) dimensionality reduction based on the number of latent states that reduces the size and complexity of the raw data [24]. To address the need for a more direct and specialized data-based optimization, while maintaining the interpretability of the derived policies, we propose an unsupervised hierarchical modeling technique that combines a high-level IOHMM with a low-level Deep Reinforcement Learning (DRL) methodology for predictive maintenance.

Hierarchical modeling is a solution towards the sample-efficient RL, which decomposes the enormous long-horizon state space into several specialized short-horizon tasks. In the first step, the IOHMM prefilters large amounts of non-relevant data generated during the normal running of the equipment and detects the state at which failure is imminent. In the second step, the DRL agent learns a policy on equipment replacement conditioned on these (close to failure) states. Our experimental results indicate that the proposed state-/event-based approach with dynamic data pre-filtering has comparable *performance*[1] to prior work that trains RL agents directly on the full data set, hence increasing the training efficiency. Lastly, it allows for more explicit interpretability of the derived policies

---

[1] performance indicates the ability to suggest replacement before failure with the use of the maximum usable life as well as with the least number of failed equipment.

by learning the latent state space. Specifically, the IOHMM learns the hidden state representation of the system (x_t) and the DRL constructs the state-action pair modeling of the environment (s_t, a_t).

To evaluate our approach, we use the NASA Commercial Modular Aero-Propulsion System Simulation (C-MAPSS), turbofan degradation data sets [4, 18]. These data sets record the output from several engine units with multivariate time series sensor readings and operating conditions discretized based on the flight cycles within a run-to-failure simulation. The following subsets of these data sets will be used in this paper: **FD001** with 1 operating condition and 1 failure mode; **FD002** with 6 operating conditions and 1 failure mode; **FD003** with 1 operating condition and 2 failure modes; and, **DS01** with ground truth degradation values.

Structure: Section 2 provides the literature review. Section 3 frames predictive maintenance as an RL problem. Section 4 proposes the novel methodology. Section 5 explains the experimental setup. Section 6 provides the interpretability aspect of the proposed methodology. Finally, Sect. 7 compares the proposed architecture with the baseline and previous work.

2   Related Work

There have been several RL methodologies developed to optimize maintenance decisions. For this task, the effectiveness of an explainable adaptive event-driven RL strategy is shown in [13,15,16] where such agents can be deployed under situation-dependent adaptations. RL in industrial applications as a predictive maintenance strategy is shown in [11,14] where the model learns from both its own experience through environment interaction as well as from the human experience feedback. The work reported in [14,21] used turbofan engines [18] as their case study for optimal maintenance decisions and discussed the limitations of prior work. In particular, they highlight that prior work is often limited to estimating the RUL of a system, giving no cause-effect relationship between the failure and the components of the equipment.

In this paper, we take the *Bayesian particle filtering* approach (Monte Carlo simulation combined with DRL) proposed in [5] as the representative of the state-of-the-art DRL for industrial maintenance and use it as a benchmark for our work. In this benchmark methodology, sequential Monte Carlo simulation is used to map the raw sensor data into latent belief degradation states [21], and it is over these latent belief states (rather than the raw sensor data) that the deep reinforcement agent learns a policy for equipment maintenance.

Given the need for interpretable decisions, researchers have also investigated the use of the Hidden Markov Model (HMM) for predicting the RUL of turbofan engines. Recent research has demonstrated the effectiveness of HMMs both towards the interpretation of fault points in terms of a correlation between a sudden decrease in RUL and transition of HMM state, as well as in terms of predicting a failure event and degradation path [8,9]. In addition, the effectiveness of Input-Output HMMs (IOHMMs), which are a more generalized version of HMM, has been explored for the diagnosis of failure, prognosis, health status, and monitoring of RUL of industrial components [10,19]. The effectiveness of online HMM estimation-based Q learning that converges to a higher mean reward for the *Partially Observable Markov Decision Process (POMDP)*, where certain variables are hidden (not directly observable), is mathematically proven by [25].

Literature Gap and Research Contributions: The majority of the research on predictive maintenance using RL focuses on the prognosis based on the estimation of RUL from multivariate raw sensor readings. However, the interpretability of the faults of the machine (at the equipment level) is missing. Furthermore, realistic environments often have partial observability, where learning from raw data might lead to suboptimal decisions. Additionally, RL encounters learning inefficiency when trained with limited samples and in an online setting [7]. In this paper, a novel methodology is proposed for maintenance decisions and interpretability that is based on DRL. At a high level, an IOHMM is designed for detecting imminent-to-failure states, while at a low level, a DRL is designed for optimizing the optimal replacement policy. Furthermore, we present a comparative analysis with prior work that demonstrates the effectiveness of the proposed methodology in terms of both performance and interpretation.

3   Framing Predictive Maintenance as an RL Problem

In this section, the decision-making problem associated with optimal predictive maintenance is framed as an RL problem.

3.1   Environment Dynamics and Modeling

The DRL framework for predictive maintenance proposed in [14] considers three actions as a general methodology for any decision-making maintenance model; *hold* [2], *repair,* and *replace.* The constraints can be the maintenance budget, and the objective function can be the maximum uptime of the equipment. We propose a general framework for modeling such environments with state transitions based on the actions selected under stochastic events (uncertainty of failure, and randomness of replacement by new equipment) at any state, as illustrated in Fig. 1. Although the general framework presented in Fig. 1 includes three actions (hold, replace, and repair), the data sets used in the experiments reported in this paper do not include data on repair actions and so for these experiments, the action space consists of just two actions (hold or replace).

---

[2]   The action of hold means that the agent neither suggests to replace nor repair and the system is healthy enough for the next operating cycle.

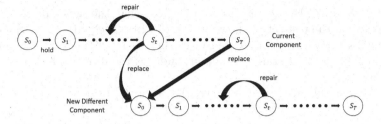

Fig. 1. Dynamics of the model of the environment

3.2 Reward Formulation

For the maintenance decision having only replacement or hold actions, a dynamic reward structure has been formulated as shown in Eq. (1) from [21]. In this equation c_r is the replacement cost, c_f is the failure cost, t is the current cycle, T_j is the final (failure) cycle, and r_t is the immediate reward.

$$
r_t = \begin{cases}
0, & a_t = \text{Hold} & \& \; t < T_j, \\
-\frac{c_r}{t}, & a_t = \text{Replace} & \& \; t < T_j, \\
-\frac{c_r + c_f}{T_j}, & a_t = \text{Hold} & \& \; t = T_j, \\
-\frac{c_r + c_f}{T_j}, & a_t = \text{Replace} & \& \; t = T_j.
\end{cases}
\tag{1}
$$

3.3 Evaluation Criteria

To evaluate the performance of the RL agent, these numerical values were chosen:

Cost. The average optimal total return $(\widetilde{Q^*})$ serves as a numeric value used and compared with the upper and lower bounds of cost for such conditions [21].

Ideal Maintenance Cost (IMC) serves as the lower bound and the ideal cost in such maintenance applications. It is the incurred cost when the replacement action is performed one cycle before the failure, as shown in Eq. (2). In this equation N denotes the number of equipment used for evaluation, $\mathbb{E}(T)$ is the expected failure state of the equipment.

$$
\phi_{IMC} \approx \frac{N \cdot c_r}{N \cdot (\mathbb{E}(T) - 1)} \approx \frac{N \cdot c_r}{\sum_{j=1}^{N} (T_j - 1)}
\tag{2}
$$

Corrective Maintenance Cost (CMC) serves as the upper bound and the maximum cost in such maintenance applications. It is the incurred cost when the replacement action is performed after the equipment has failed as shown in Eq. (3).

$$
\phi_{CMC} \approx \frac{(c_r + c_f)}{\mathbb{E}(T)} \approx \frac{N \cdot (c_r + c_f)}{\sum_{j=1}^{N} T_j}
\tag{3}
$$

Average Optimal Cost ($\widetilde{Q^*}$) is the average cost that the agent receives as its performance on the test set as shown in Eq. (4). In this equation $r(s,a)$ denotes the immediate reward as formulated in Eq. (1), $Q^*(s',a')$ denotes the optimal action value of the next state-action pair, and γ is the discount factor.

$$\widetilde{Q^*}(s,a) = \frac{1}{N}\sum \left[r(s,a) + \gamma \max_{a'} Q^*\left(s',a'\right) \right] \tag{4}$$

Average Remaining Useful Life ($\widetilde{RUL}$) before replacement. It quantifies; how many useful cycles are remaining on average when the agent proposes the replacement action. Ideally, it should be one according to our defined criteria.

4 Proposed Methodology (SRLA)

The proposed hierarchical methodology integrates an IOHMM and a DRL agent. Within this hierarchical model, the purpose of the IOHMM is to identify when the system is approaching a desired (in our case: failure) state. Once the IOHMM has reached this failure state, the DRL agent's task is to optimize the decision on when to replace the equipment to maximize its total useful life. This IOHMM-DRL model allows for state- or event-based optimization. This further allows for a more efficient DRL training, since the training data set is restricted to the imminent-to-failure states. Figure 2 illustrates the proposed model which we name Specialized Reinforcement Learning Agent (SRLA).

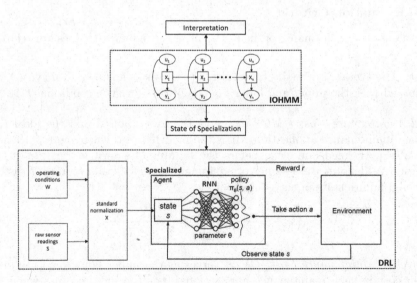

Fig. 2. Specialized Reinforcement Learning Agent (SRLA).

The DRL training and optimization process is relatively standard. We use Deep Learning (DL) as a function approximator that generalizes effectively to enormous state-action spaces through the approximation of unvisited states [3] as shown in Eq. (5). In this equation L_i denotes the loss function, y_i is the TD target; which is the sum of the observed one-step reward and the discounted next Q (action) value conditioned on the current state and action, $Q(s, a)$ is the estimation of the Q value of the current state-action pair parameterized by θ.

$$L_i(\theta_i) = \mathbb{E}_{a \sim \mu} \left[(y_i - Q(s, a; \theta_i))^2 \right];$$

$$y_i := \mathbb{E}_{a' \sim \pi} \left[r + \gamma \max_{a'} Q(s', a'; \theta_{i-1}) \mid S_t = s, A_t = a \right] \tag{5}$$

At a high level, an IOHMM is used, where the objective of training optimization is to identify the model parameters that best determine the given sequence of observations conditioned on the given input. In the context of industrial settings, these inputs are the operating conditions that heavily influence the state of the system and control the system's behavior. Parameter γ is the vector defining the probability of being in each hidden state at a particular time $x_t = S_i$; given the input U, the observation sequence Y, and the parameters of the trained model λ (initial state, transition, and emission probability matrices conditioned on the input (U) as well), as shown in Eq. (6). Parameter δ from Eq. (7) in this context is used to predict the health degradation state sequence of the equipment, where the last cycle of each equipment determines the failure state. The inference algorithm for the SRLA is described in Algorithm A.1 of Appendix A.

$$\gamma_t(i) = P(x_t = S_i \mid U, Y, \lambda) \tag{6}$$

$$\delta_t(i) = \max_{x_1, \cdots, x_{t-1}} P[x_1 \cdots x_t = i, Y_1 \cdots Y_t \mid U, \lambda] \tag{7}$$

4.1 Interpretability with IOHMM

Beyond the performance considerations of the model, the IOHMM component provides a level of interpretability in terms of identifying failure states, the root cause of failure, and stages of health degradation. Based on the state sequence distributions predicted by the IOHMM from Eq. (7), each state of a particular event can be decoded, such as the failure mode or degradation stage, as shown in [8]. To discover the most relevant sensor readings corresponding to these failure states that triggered the IOHMM to predict such a state, feature importance is performed that leads to the root cause failure analysis. Raw sensor readings are used as the input feature for the model and IOHMM state predictions are used as the target. After fitting the model, the importance of each sensor can be extracted for each IOHMM state. Apart from the failure event hypothesis, it is necessary to measure the health state of the equipment at different points to generate an alarm for the user when the equipment reaches a critical point of its lifetime. The interpretations are based on the critical points along the equipment degradation curve as shown in Fig. 3 and the range of observed IOHMM states.

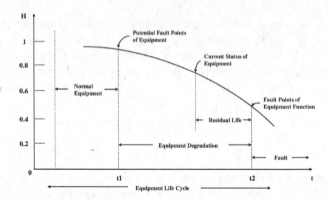

Fig. 3. Health degradation curve of equipment, taken from [12]

5   Experimental Setup

The two baseline systems defined in this paper are distinguished and designed by varying each of these four stages: (i) input, (ii) feature engineering, (iii) RL architecture, and (iv) output. The summary of the training parameters is shown in Appendix A.1 of Appendix A.

Baseline 1: Sensor Data + Operating Conditions: (i) Raw sensor data and operating conditions as the input, (ii) Standard normalization as the feature extraction module, (iii) DNN as the RL architecture, and (iv) Action policy at the output. It is used to set the failure cost to be used for the rest of the experiments.

Baseline 2: Sensor Data + Operating Conditions + IOHMM: (i) Raw sensor data and operating conditions as input, (ii) MinMax normalization and IOHMM as the feature engineering module, (iii) RNN as RL architecture, and (iv) Action policy, RUL estimation, and unsupervised clustering and interpretation based on events at output; as shown in Fig. 4. Its significance is to determine the optimal number of IOHMM states to be used in the experiments. Implementation of IOHMM is done through a library [23]. This baseline uses the output of the IOHMM (probability distribution) as the input to the DRL agent, whereas SRLA uses the raw data as the input to the DRL agent during the state of specialization.

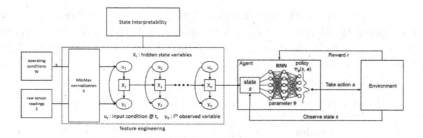

Fig. 4. IOHMM posterior probabilities as the input to DRL.

5.1 Setting the Hyperparameters for the Models

This section describes the experiments used to determine the hyperparameters
(i) cost of failure (c_f) and (ii) number of IOHMM states. The effectiveness of
the architectures has been evaluated as described in Sect. 3.3. The data set used
for this part of the experiment is FD001, which is split into an 80:20 (train:test)
ratio.

Calculating the Cost of Failure. The reward function (Eq. (1)) for the RL
agent requires the specification of *cost of failure* (c_f) and *cost of replacement*
(c_r). However, the NASA C-MAPSS data set does not specify these parameters.
To fix these values, we train Baseline 1 using a range of different c_f, while fixing
c_r and then comparing and identifying the c_f that minimizes the average of
total optimal cost per episode ($\widetilde{Q^*}$). c_r is fixed (100) and the comparison is
based on the different c_f values of 25, 500 and 1000, as shown in Table 1. It
was observed that as c_f increases, $\widetilde{Q^*}$ becomes closer to the ideal cost, and, at
the same time, the number of failed units decreases to 0%. However, the agent
becomes more cautious, suggesting replacement action earlier in the lifetime of
the engine; thereby, increasing the average remaining cycles. In the context of
predictive maintenance of safety-critical systems, it is more important to avoid
failure at the expense of replacing equipment a few cycles before its remaining
useful life. Therefore, c_f of 1000 was chosen for the rest of the experiments.

Calculating the Number of Hidden States. Baseline 2 was used to find
the number of states of the IOHMM model that maximizes the likelihood of our
state space and the performance of the DRL through an iterative process. We
evaluated the performance of the model as the number of states varied between
10, 15, and 20 states. The model trained through IOHMM gives the posterior
probability distribution for every state as shown in Eq. (6), which is then fed
as an input to the DRL agent to be able to learn the optimal maintenance
(replacement) policy. The experiment was carried out on the test set using the

failure cost of 1000 and with the same parameters as the previous experiment for a better evaluation. 15 states of the IOHMM showed better performance results than the rest, and so in the rest of our experiment, we use 15 as the number of states for IOHMM model.

Table 1. Comparative evaluation and hyperparameter search.

Failure cost	Avg Q^*	IMC	CMC	$\widetilde{RUL}$	Failed units
Baseline 1					
25	0.54	0.45	0.56	2.4	45%
500	0.61	0.45	2.68	7.5	5%
1000	0.49	0.45	4.92	7.0	0%
Baseline 2					
IOHMM states					
10	0.54	0.45	4.92	24.2	0%
15	0.49	0.45	4.92	6.8	0%
20	0.53	0.45	4.92	20.2	0%

6   Experiment 1: Interpretations Based on Hidden States

Data sets FD001, FD003, and DS01 are used in this section using the IOHMM for event-based hypothesis and state interpretations. The experiments performed here are to address the question of whether the introduction of the hidden states can help towards interpretability.

6.1   Interpretability - Failure Event Hypothesis

Due to the unavailability of the ground truth for other state mappings in FD003, just the failure states (last cycle state) were mapped in this experiment. Each failure state in the dataset is annotated with one of the 2 failure modes (HPC and fan degradation); however, the ground truth for the engines corresponding to which failure mode is not provided. Analyzing the failure states revealed two IOHMM states that corresponded to the failure event, which might be based on the two failure modes. To validate this hypothesis, the analysis was repeated with FD001, where there is only one failure mode defined in the description of the data set, and this analysis showed that only one IOHMM state was observed to be the failure state for each engine. This suggests that it is possible to map IOHMM states to failure events within the health state of the equipment.

Using the feature importance methodology described in Sect. 4.1, features (sensor readings) with a relatively higher score (based on feature importance) were selected from each class (failure states depicted by IOHMM). Further, the corresponding actual sensor information and description were extracted from [18] as described in Table 2. From the background information from the sensor descriptions, it was observed that the sensor importance for the two different IOHMM states showed a concrete failure event interpretation that corresponded to the failure described in the data set (HPC and Fan degradation), as hypothesized in Table 3.

Table 2. Feature to sensor description.

Feature	Sensor	Description
5	P_{30}	HPC outlet pressure
8	epr	Engine pressure ratio
10	phi	fuel flow : HPC pressure
13	BPR	Bypass ratio

Table 3. Sensor importance to failure event.

IOHMM state	Important sensor reading	Failure event hypothesis (interpretation)
9	BPR	Fan degradation
14	P_{30}, epr, phi	HPC degradation

6.2 Interpretability - State Decoding and Mapping

The second version of the NASA C-MAPSS data set [4] was used here to evaluate the state interpretability of IOHMM throughout the engine life, a subset of which is shown in Fig. 5, where the red trend represents the IOHMM state prediction based on Eq. (7). The data set has the ground truth values of the engine's state per cycle, the Boolean health state value is represented by the blue line with state 1 being healthy and 0 being unhealthy, the RUL is represented by the yellow line, and the green curve represents the health degradation curve. Based on the reference health degradation curve from Fig. 3, and the range of IOHMM states observed during those conditions, we were able to associate different IOHMM states with different equipment conditions as shown in Table 4.

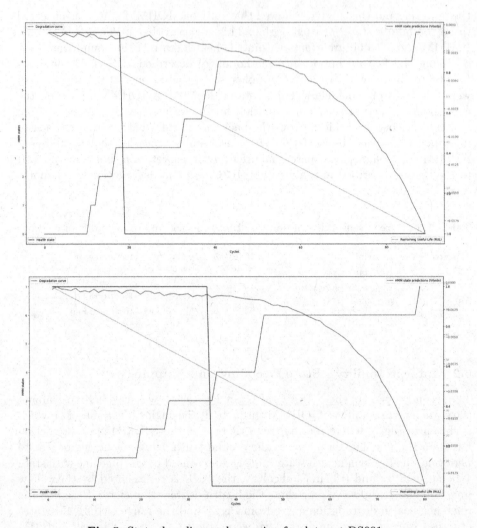

Fig. 5. State decoding and mapping for data set DS001.

Table 4. Interpretability of the IOHMM state to equipment conditions.

Equipment condition	IOHMM states
Normal equipment	0–2
Potential fault point of equipment	2–4
Failure progression	4–6
Fault point of equipment function	6–7
Failure	7

7 Experiment 2: Comparison of SRLA with Prior Work

Data set FD002 is used in this experiment for the comparative evaluation with baselines and prior work [21].

7.1 Comparative Evaluation and Results

As seen in Sect. 6.2, the IOHMM can align its states and state transitions with the relevant health states of the engine; however, the definition and alignment of the states were not fine enough to replace the engine with just one cycle before the failure. Therefore, DRL is used to refine the granularity after state distribution based on IOHMM, resulting in a hierarchical model. To evaluate the performance, the results are compared with the two baseline systems and the Particle Filtering (PF) based-DRL (*Bayesian particle filtering*) framework proposed by [5]. In their experiments [5] used 80 engines as the training set and 20 as the test set out of 260 engines. However, the engines were selected randomly; therefore, an exact comparison with the average agent cost could not be made. Therefore, the ratio of the Ideal Maintenance Cost (IMC) to the average agent cost ($\widetilde{Q^*}$) was compared in Table 5. As shown, Baseline 2 performs better than Baseline 1 and SRLA outperforms baseline systems and has a comparative performance with the PF + DRL methodology with the added benefits of interpretability.

Table 5. Comparison of the proposed methodology with baseline systems and [21].

Methodology	$\widetilde{Q^*}$	IMC	CMC	IMC/$\widetilde{Q^*}$	Failure	$\widetilde{RUL}$	Interpretations
Baseline 1	6.87	0.64	7.02	0.09	90%	2.6	No
Baseline 2	0.77	0.64	7.02	0.83	0%	23.0	Yes
PF + DRL [15]	2.02	1.93	20.80	0.96	0%	-	No
SRLA	0.69	0.64	7.02	0.94	0%	6.4	Yes

8 Conclusion and Future Direction

In this paper, a new hierarchical methodology was proposed utilizing the hidden Markov model-based deep reinforcement learning allowing the functionality of interpretability in the stochastic environment along with defining an optimal replacement policy and estimating remaining useful life without supervised annotations. Therefore, such a model can easily be used in industrial cases where the annotation of the fault type is difficult to obtain and the human supervisor in the loop can help define the state distribution according to the event-based analysis. To test the effectiveness of the model, the NASA C-MAPSS (turbofan engines) data sets versions 1 and 2 were used. It was compared with baseline models and prior work of Bayesian filtering-based-deep reinforcement learning to evaluate the performance. Our results indicate that the IOHMM-DRL framework outperforms the baseline DRL systems and has performance comparable to the Bayesian filtering DRL approach, with the added benefits of interpretability and a less complex system model. In the future, the proposed hierarchical architecture of IOHMM-DRL will be applied to other open data sets along with real-world case studies to measure its robustness.

A   Algorithms and Training Parameters

Algorithm A.1. Specialized Reinforcement Learning Agent (SRLA)

STEP I: IOHMM Training
Input:
n: number of hidden states
Y: output sequences
U: input seauences
Output: λ: model parameters (initial, transition, and emission probability)

STEP II: Viterbi Algorithm (IOHMM Inference)
Input: λ, U, Y
Output: $\delta_t(i) = \max_{x_1, \cdots, x_{t-1}} P\left[x_1 \cdots x_t = i, u_1 \cdots u_t, y_1 \cdots y_t \mid \lambda\right]$

STEP III: DRL Training
Input:
δ_s: specific event (such as failure)
S_t: $u_t + y_t$
Enviroment Modeling
Deep Reinforcement Learning
Output: $\hat{Q}^*(S_t, A_t)$

STEP IV: SRLA Inference
Input: λ, $\hat{Q}^*(S_t, A_t)$, S_t: (U_t, Y_t)
Step II, Interpretations Based on Hidden States
$\delta \rightarrow$ Specialized state $(X_s) \rightarrow U_s, Y_s$
if S_t in X_s **then**
   $\hat{Q}^*(s_t, a_t)$
   Environment Model
else
   a_t = do nothing (hold)
end if
Output: $\hat{Q}^*(\delta_t, s_t, a_t)$

A.1   Training Parameters

The summary of the DL framework within the RL architectures is as follows: (a) Deep Neural Network (DNN) consisting of a total of 37,000 training parameters and fully-connected (dense) layers with 2 hidden layers that have 128 and 256 neurons, respectively, with ReLU activation. (b) Recurrent Neural Network (RNN) consists of 468,000 training parameters and fully connected (LSTM) layers with 2 hidden layers having 128 and 256 neurons, respectively. The output layer consists of the number of actions the agent can decide for decision-making with linear activation. The parameters of the DRL agent are as follows: discount rate = 0.95, learning rate = 1e−4, and the epsilon decay rate = 0.99 is selected with the initial epsilon = 0.5.

References

1. Bengio, Y., Frasconi, P.: Input-output hmms for sequence processing. IEEE Trans. Neural Netw. **7**(5), 1231–1249 (1996). https://doi.org/10.1109/72.536317
2. Bengio, Y., Frasconi, P.: An input output hmm architecture. In: Advances in Neural Information Processing Systems, pp. 427–434 (1995)
3. Bertsekas, D.P., Tsitsiklis, J.N.: Neuro-dynamic programming. In: Athena Scientific (1996)
4. Chao, A., et al.: Aircraft engine run-to-failure dataset under real flight conditions for prognostics and diagnostics. Data **6**(1), 5 (2021)
5. Chen, Z., et al.: Bayesian filtering: from Kalman filters to particle filters, and beyond. Statistics **182**(1), 1–69 (2003)
6. Do, P., et al.: A proactive condition-based maintenance strategy with both perfect and imperfect maintenance actions. Reliab. Eng. Syst. Saf. **133**, 22–32 (2015)
7. Dulac-Arnold, G., et al.: Challenges of real-world reinforcement learning: definitions, benchmarks and analysis. Mach. Learn. **110**(9), 2419–2468 (2021). https://doi.org/10.1007/s10994-021-05961-4
8. Giantomassi, A., et al.: Hidden Markov model for health estimation and prognosis of turbofan engines. In: International Design Engineering Technical Conferences and Computers and Information in Engineering Conference, vol. 5480 (2011)
9. Hofmann, P., Tashman, Z.: Hidden markov models and their application for predicting failure events. In: Krzhizhanovskaya, V.V. (ed.) ICCS 2020. LNCS, vol. 12139, pp. 464–477. Springer, Cham (2020). https://doi.org/10.1007/978-3-030-50420-5_35
10. Klingelschmidt, T., Weber, P., Simon, C., Theilliol, D., Peysson, F.: Fault diagnosis and prognosis by using input-output hidden markov models applied to a diesel generator. In: 2017 25th Mediterranean Conference on Control and Automation (MED), pp. 1326–1331 (2017). https://doi.org/10.1109/MED.2017.7984302
11. Lepenioti, K., et al.: Machine learning for predictive and prescriptive analytics of operational data in smart manufacturing. In: Dupuy-Chessa, S., Proper, H.A. (eds.) CAiSE 2020. LNBIP, vol. 382, pp. 5–16. Springer, Cham (2020). https://doi.org/10.1007/978-3-030-49165-9_1
12. Li, H.Y., Xu, W., Cui, Y., Wang, Z., Xiao, M., Sun, Z.X.: Preventive maintenance decision model of urban transportation system equipment based on multi-control units. IEEE Access **8**, 15851–15869 (2019)

13. Meng, F., An, A., Li, E., Yang, S.: Adaptive event-based reinforcement learning control. In: 2019 Chinese Control And Decision Conference (CCDC), pp. 3471–3476. IEEE (2019)
14. Ong, K.S.H., Niyato, D., Yuen, C.: Predictive maintenance for edge-based sensor networks: a deep reinforcement learning approach. In: 2020 IEEE 6th World Forum on Internet of Things (WF-IoT), pp. 1–6. IEEE (2020)
15. Panzer, M., Bender, B.: Deep reinforcement learning in production systems: a systematic literature review. Int. J. Prod. Res. 1–26 (2021)
16. Parra-Ullauri, J.M., et al.: Event-driven temporal models for explanations - ETe-MoX: explaining reinforcement learning. Softw. Syst. Model. **21**(3), 1091–1113 (2021). https://doi.org/10.1007/s10270-021-00952-4
17. Rabiner, L., Juang, B.: An introduction to hidden markov models. IEEE ASSP Mag. **3**(1), 4–16 (1986). https://doi.org/10.1109/MASSP.1986.1165342
18. Saxena, A., Goebel, K.: Turbofan engine degradation simulation data set. In: NASA Ames Prognostics Data Repository, pp. 878–887 (2008)
19. Shahin, K.I., Simon, C., Weber, P.: Estimating iohmm parameters to compute remaining useful life of system. In: Proceedings of the 29th European Safety and Reliability Conference, Hannover, Germany, pp. 22–26 (2019)
20. Sikorska, J., Hodkiewicz, M., Ma, L.: Prognostic modelling options for remaining useful life estimation by industry. Mech. Syst. Sig. Process. **25**(5), 1803–1836 (2011)
21. Skordilis, E., Moghaddass, R.: A deep reinforcement learning approach for real-time sensor-driven decision making and predictive analytics. Comput. Ind. Eng. **147**, 106600 (2020)
22. Sutton, R.S., Barto, A.G.: Reinforcement Learning: An Introduction. MIT press (2018)
23. Yin, M., Silva, T.: Iohmm (2017). https://github.com/Mogeng/IOHMM
24. Yoon, H.J., Lee, D., Hovakimyan, N.: Hidden markov model estimation-based q-learning for partially observable markov decision process. In: 2019 American Control Conference (ACC) (2019). https://doi.org/10.23919/acc.2019.8814849
25. Yoon, H.J., Lee, D., Hovakimyan, N.: Hidden markov model estimation-based q-learning for partially observable markov decision process. In: 2019 American Control Conference (ACC), pp. 2366–2371. IEEE (2019)

Pathology Data Prioritisation: A Study Using Multi-variate Time Series

Jing Qi[1(✉)], Girvan Burnside[2], and Frans Coenen[1]

[1] Department of Computer Science, The University of Liverpool,
Liverpool L69 3BX, UK
j.qi7@liverpool.ac.uk
[2] Department of Health Data Science, Institute of Population Health,
The University of Liverpool, Liverpool L69 3BX, UK

Abstract. A mechanism to support the prioritisation of multi-variate pathology data, in the absence of a ground truth prioritisation, is presented. The motivation is the ever increasing quantity of pathology data that clinicians are expected to consider. The fundamental idea, given a previously unseen pathology result and the associated pathology history, is to use a deep learning model to predict future pathology results and then use the prediction to classify the new pathology result according to a pre-defined set of prioritisation levels. A further challenge is that patient pathology history, expressed as a multi-variate time series, tends to be irregularly time stamped and of variable length. The proposed approach used a Recurrent Neural Network to make predictions and a bounding box technique for the classification. The approach was evaluated using Urea and Electrolytes pathology data. The operation of the proposed approach was also compared with previously reported approaches, and was found to outperform these previous approaches.

Keywords: Data ranking · Multivariate time series · Deep learning · Pathology data

1 Introduction

Pathology results play an important role for decision making in any clinical environment. Clinicians use pathology results to diagnose patient conditions and decide on best next measures. Large amounts of pathology results are generated on a daily basis. For many conditions, such as kidney disease, pathology results are generated at regular intervals, sometimes over many years. Many pathology results comprise a set of values, not just one; in other words they are multivariate. The amount of pathology data that clinicians are expected to look at on a daily bases presents a significant information overload problem. A problem that is compounded by our ever increasing technical ability to collect pathology data; not helped by the recent COVID-19 pandemic which has put further strain on resources. In order to solve the problem, it is suggested that some form of

R. Wrembel et al. (Eds.): DaWaK 2022, LNCS 13428, pp. 149–162, 2022.
https://doi.org/10.1007/978-3-031-12670-3_13

automated pathology result prioritisation is required, and that this can best be achieved using the tools and techniques of machine learning whereby results can be classified using a prioritisation scale of some kind. However, the challenge of the application of deep learning to pathology data is the absence of a "ground truth", a set of examples illustrating what a priority pathology result looks like, and what it does not look like. The reason for this is the resource required to generate such a ground truth.

There has been some previous work directed a pathology result prioritisation in the absence of a ground truth [7,8]. In [8] it was assumed that high priority pathology results equated to anomalous priority results and hence an anomaly detection mechanism was adopted. However, given a large number of priority pathology results these would no longer be considered to be anomalous and therefore not be prioritised. In [7] a proxy ground truth was used based on the known outcomes of previous patients; whether they became emergency patients, in-patients, out-patents or remained with their General Practitioner (GP). The proxy ground truth was used to train a deep learner. Some improvement was reported over the work presented in [8]. However, the way the proxy ground truth was calculated meant that possible correlations between different pathology values were not considered.

An alternative pathology result prioritisation mechanism, to that given in [7] and [8], is presented in this paper directed at patients that have conditions where pathology results, each comprised of a set of values, are generated as an ongoing part of a care programme. In other words we have time series of previous pathology results. The fundamental idea is to predict whether the next pathology result in the sequence will be out of the anticipated normal range with respect to the pathology test under consideration. To be more specific, the use of an RNN-based pathology result forecasting model is advocated to predict follow-on results which can then be compared to the expected range. To distinguish prioritisation levels, a novel "Bounding Box" mechanism, which can distinguish between levels of prioritisation, is proposed. For the evaluation presented later in this paper three prioritisation levels were considered: high, medium and low. However, the bounding box technique will also work with any numbers of levels of two or more.

From the foregoing, the hypothesis that this paper seeks to establish is that there are patterns (trends) in a patients' historical pathology data which can be considered to be markers that are indicative of future pathology values. To act as a focus for the work the application domain of Urea and Electrolytes (U&E) pathology testing was considered. The proposed approach was evaluated using U&E data provided by Arrowe Park Hospital in Merseyside in the UK. This application domain was selected because it is the most commonly undertaken biochemistry test used to provide essential information on renal function.

The remainder of this paper is organised as follows. A review of relevant previous work is presented in Sect. 2. This is followed, in Sect. 3, by a review of the Urea and Electrolytes pathology application domain, used as a focus for the work. The proposed approach is considered in Sect. 4, and the evaluation of the proposed approach in Sect. 5. The paper is concluded in Sect. 6 with a summary of the main findings and some suggested directions for future work.

2 Previous Work

The broad area of research into which the work presented in this paper falls, is that of big data prioritisation [12], where the aim is to determine which data items take priority over other data items. Data prioritisation can be applied in many areas [1,6,10], while in the medical area, the concept is more similar to patient triage [3] or prognosis [2], which supports decision-making through predicting the severity or risk of a given patient's condition. There has been some previous work directed at using machine learning for patient triage [4, 5,14]. In [4] various forms of multinomial Logistic Regression (LR) were used: multinomial LR, eXtreme gradient boosting (XGBoost), random forests (RFs) and Gradient-Boosted Decision Trees (GBDTs) were explored to identify high-risk emergency department patients with suspected cardiovascular disease. In [5], natural language prediction methods were adopted to predict admission to a Neurosciences intensive Care Unit. In [14] a machine Learning based AutoScore-Derived triage tool was developed for predicting mortality risk after patients admitted to emergency departments. Most of the work aimed at predicting triage adopted supervised learning using a predefined training data labeled by domain experts, a resource intensive process which does not scale up to give general applicability. One of the challenges for the prioritisation of pathology data is the absence of training data.

Another challenge of time series pathology data is that it is usually irregular [15]; the spacing of observations is not constant. Most time series prediction methods, the technology adopted with respect to the approach presented in this paper, assume unit-spaced (regular) time series data [11]. Pathology time series data also tends to be multivariate in nature; the time series have more than one time-dependent variable. Each value depends not only on its past values but also the values for the associated variables. There are very few reported studies where irregular time series have been used directly. The majority of studies adopt some form of imputation so that spacing is of a unit length to ensure that time series are all of the same length. Or alternatively some form of padding and masking is used [13]. A number of alternatives are considered later in this paper.

3 U&E Testing Application Domain

The work presented in this paper is focused on Urea and Electrolytes pathology test data, U&E testing. U&E testing is usually performed to confirm normal kidney function or to exclude a serious imbalance of biochemical salts in the bloodstream. The U&E test data considered in this paper comprised five values per record: (i) Bicarbonate (bi), (ii) Creatinine (cr), (iii) Potassium (po), (iv) Sodium (so) and (v) Urea (ur). The measurement of each is referred to as a "task", thus we have five tasks per test. Abnormal levels in any of these tasks may indicate that the kidneys are not working properly. However, a one time abnormal result does not necessarily need to be prioritised. A new task value that is out of range for a patient who has a previous recent history of out of range task values, but the latest result indicates a trend back into the normal range, may not be a priority result. Conversely, a new task value that is within the normal range for a patient who has a history of normal range task values, but the latest result indicates a trend heading out of the normal range, may need to be prioritised. Given a new set of pathology values for a U&E test we wish to determine the priority to be associated with this set of values.

The U&E data used for evaluation purposes with respect to the work presented in this paper comprised a set of clinical patient records, $\mathbf{D} = \{P_1, P_2, \ldots\}$, where each record $P_i \in \mathbf{D}$ was of the form:

$$P_i = \langle PatientID, History, TestResult, ReferencedRange \rangle \qquad (1)$$

where: (i) $PatientID$ is the ID for the patient in question; (ii) $History$ is a set previously obtained pathology results expressed as a set a multivariate time series $T = [t_1, t_2, \ldots]$, where each $t_i \in T$ comprised a 5-tuple of the form $\langle v_{bi}, v_{cr}, v_{po}, v_{so}, v_{ur} \rangle$, (iii) $TestResult$ is a current previously unseen pathology result R also comprised of an n tuple of the form $\langle v_{bi}, v_{cr}, v_{po}, v_{so}, v_{ur} \rangle$, and (iv) $ReferencedRange$ is a set of bounds defining the normal range for the patient in question for the values associated with each task represented as two sets, $L = [l_1, l_2, \ldots]$ and $U [u_1, u_2, \ldots]$, where L holds the minimum (normal low) values and U holds the maximum (normal upper) values. There is a one-to-one correspondence with T. The normal low and high dimensions indicate a "band" in which pathology results are expected to fall given a healthy patient. These bands vary from task to task, will not be the same for each patient and may change for a given patient over the course of time. A training record $P_i \in \mathbf{D}$ will also include a task label c taken from a set of classes C.

4   Prioritisation for U&E Pathology Patients Results

The fundamental idea, for prioritising pathology results, presented in this paper is that prioritisation can be achieved by predicting whether future patient pathology results will be out of the normal range. An overview of the proposed process is given in Fig. 1. The process commences with a new multi-variate pathology result $R = \langle v_1, v_2, \ldots, v_n \rangle$. This is combined with the pathology history of the patient in question to give a time series $T = [t_1, t_2, \ldots, R]$. This is then passed to a prediction model where the next set of results $P = \langle v_1, v_2, \ldots, v_n \rangle$ is predicted. A bounding-box technique is then used to classify R according to P. Thus, Fig. 1, there are two main stages within the overall process:

1. **Future Result Prediction:** The process of predicting future pathology results given a new, previously unseen, multi-variate pathology result and the pathology result history for a given patient.

2. **Bounding Box Classification:** The process of assigning a priority level to a previously unseen pathology result using the predicted future pathology results.

Each of the above is discussed in further detail in the following two sub-sections.

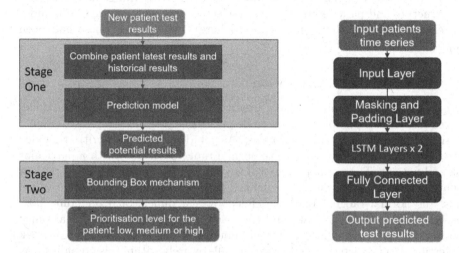

Fig. 1. Schematic outlining the proposed pathology data prioritisation process

Fig. 2. LSTM architecture for future pathology results prediction

4.1   Future Results Prediction

There are a range of prediction mechanism that could have been adopted with respect to Stage 1 of the proposed process, a Long Short Term Memory (LSTM) mechanism was adopted. The architecture of the proposed LSTM is showed in Fig. 2. From the figure it can be seen that the proposed LSTM comprised four layers: (i) the input layer which receives a time series T which includes the new pathology result R of interest, (ii) the padding layer where the time series were transformed from irregular time series to regular time series, (iii) the hidden layer comprised of LSTM cells arranged in two sub-layers, and (iv) the fully connected layer where the predicted future test results are generated. The hidden layer is where the training takes place. The training set is divided into a set of overlapping input/output *samples*. During training, each sample forms a *prediction step* in the overall LSTM model generation process. The hidden layers compute the intermediate results and pass them on to the next iteration, which makes it possible for the network to maintain memory of the state of earlier historical records, so that the effect of the early results can be considered for the prediction.

The adopted LSTM architecture comprised 16 cells arranged in two sub-layers in the hidden layer, and n neurons in the output layer for predicting a set of n pathology task values ($n = 5$ with respect to the U&E data set used to evaluate the proposed approach). The "input shape" of the data is a time series $T = [t_1, t_2, \ldots]$, where each point t_1 comprises a n-tuple. An important aspect of training a LSTM network is the iterative updating of weights using the training data. To achieve this, the Adam stochastic gradient descent was used. The adopted loss function was the Mean Squared Error (MSE) loss function. The structure and parameters used were selected a as consequence of a number of preliminary experiments (not reported here), and because these had been adopted in related work [9].

4.2   Bounding Box Classification

Once a set $P = \langle v_1, v_2, v_3, \ldots, v_n \rangle$ of predicted test results have been obtained, derived from a new pathology result to be prioritised, the next step is to assign a priority class to the new pathology result. As noted earlier, the prioritisation idea considered in this paper is to use the normal range associated with a particular task and patient. The normal range will vary from task to task and from patient to patient, but in each case will be defined by a minimum and maximum value. Thus, the normal range "zone" is defined by a n dimensional bounding box, where n is the number of tasks and each side will equate to the normal range associated with a task defined by the minimum and maximum value for the task in question. If a predicted pathology result falls entirely within this bounding box the pathology result will be deemed to have a "low" priority. Anything outside can then be labelled as "high" priority. However, a binary classification (high-low) is considered too coarse a classification; we require more than one class label for results that fall outside of the "low priority bounding box". In this paper we will consider a three class prioritisation, $C = \{\text{high}, \text{medium}, \text{low}\}$. Thus if a pathology result falls outside of the "low priority bounding box" it will be either medium or high priority. The question is how this can best be calculated. One idea is to simply calculate the Euclidean distance from the geometric centre of the low priority bounding box and use a threshold of some kind to distinguish high priority pathology results from low priority results. However, this will mean that the distance from a pathology result close to a corner of the bounding box to the geometric center, will be treated the same as a result some way away from a side of the bounding box. Thus to distinguish between medium priority and high priority pathology result a second bounding box, the "medium priority bounding box" was defined by expanding the low priority box by a factor χ.

The pseudo code for the Bounding Box Comparison approach is given in Algorithm 1. Note that the algorithm assumes $n = 5$. The inputs to the algorithm are: (i) a predicted pathology result $P = \{v_1, v_2, \ldots v_n\}$, (ii) the set $L = \{l_i, l_2, \ldots l_n\}$ of low normal range values, (iii) the set $U = \{u_i, u_2, \ldots u_n\}$ of upper normal range values and (iv) the expansion factor χ to be applied. The algorithm commences, line 2, by defining a default class of "high". Then, line 3, the algorithm determines whether P falls inside the low priority bounding box using a function *inNormalZone*. This returns a result tuple comprised of a set of binary values, 0 = inside and 1 = outside. In the U&E case the tuple is

Algorithm 1. Bounding Box Comparison

```
1: input P, L, U χ
2: class = high                                              ▷ Default class
3: result = inNormalZone(P, U, L)              ▷ Determine if P in normal range
4: if result == ⟨0, 0, 0, 0, 0⟩ then
5:     class = low                    ▷ Predicted point entirely within normal zone
6: else
7:     result = ⟨1, 1, 1, 1, 1⟩
8:     for ∀vᵢ ∈ P do
9:         offset = (uᵢ − lᵢ) × χ
10:        if (uᵢ + offset) ≥ tᵢ ≥ (lᵢ − offset) then
11:            resultᵢ = 0
12:        end if
13:    end for
14:    if result == ⟨0, 0, 0, 0, 0⟩ then
15:        class = medium              ▷ Predicted point entirely within medium zone
16:    end if
17: end if
18: return class
```

of size $n = 5$, but it can be any other values according to the type of pathology under consideration. If the resulting tuple equates to $\langle 0, 0, 0, 0, 0 \rangle$ then P is entirely within the low priority bounding box and allocated the class "low" (line 5). Otherwise the result tuple is set to $\langle 1, 1, 1, 1, 1 \rangle$ (line 7) and we proceed to expand the low priority bounding box in each direction in an iterative manner (lines 8 to 13). On each iteration the expansion is conducted using an offset applied to each normal low and upper value. The offset, with respect to each task is calculated as shown in Eq. 2, where $l_i \in L$, $u_i \in U$, and χ is a predefined multiplier (factor). On each iteration (line 10), $t_i \in R$ is compared with the expanded range and the outcome added to the result tuple. On completion, if the tuple equates to $\langle 0, 0, 0, 0, 0 \rangle$ the new pathology record P is allocated the class "medium" (line 15). Otherwise the default class, "high", is used. The class is then returned (line 18). It is easy to see how the process can be repeated and further classes added if desired.

$$offset = (u_i - l - i) \times \chi \qquad (2)$$

A value for chi can be established empirically. However, for the evaluation presented here a proxy ground truth was used (more on this in Subsect. 5.1). The proxy training data set was of the form presented in Subsect. 3. A value for χ was then "learnt", for each class, by clustering all potential χ values and then determining the mid point between the two cluster centroids. For the evaluation presented in the following section a binary classification scenario was considered $C = \{c_1, c_2\}$. The adopted process is illustrated in Algorithm 2. The input to the algorithm is the training set **D** and the set of classes C (see Sect. 3). A value for χ for each value v_k in each time series T_j in each set of time series T for each patient record D_i in **D** was calculated using the Equation 3 where u_k and l_k are the upper and lower range limits associated with v_k, and $dist$ is the distance of v_k from the mid-point between u_k and l_k calculated as shown in Equation 4. The average value χ for each class is calculated, lines 12 and 13 in Algorithm 2. The mean of the two averages is then the final value for χ to be used. For future work the intention is to investigate the potential of using different values for chi for different tasks. In the present study the same value of χ was used throughout.

Algorithm 2. Factor χ Generation

1: **Input D**, C
2: **Chi** = $\{Chi_1, Chi_2\}$
3: **for** $D_i \in$ **D do**
4:     **for** $T_j \in$ **T**, **T** $\in D_i$ **do**
5:         **for** $v_k \in T_j$ **do**
6:             $dist = abs\left(v_k - \frac{u_k - l_K}{2}\right)$
7:             $\chi = \frac{2 \times dist}{u_k - l_k}$
8:             $Chi_i = Chi_i \cup \chi$, $i =$ class ID for class $c \in D_i$
9:         **end for**
10:     **end for**
11: **end for**
12: $ave_{c_1} = average(Chi_1)$ ▷ cluster centre for $c_1 \in C$
13: $ave_{c_2} = average(Chi_2)$ ▷ cluster centre for $c_2 \in C$
14: $\chi = \frac{ave_{c_1} + ave_{c_2}}{2}$
15: **return** χ

$$\chi = \frac{2 \times dist}{u_k - l_k} \quad (3) \qquad dist_i = abs\left(v_k - \frac{u_k - l_K}{2}\right) \quad (4)$$

5 Evaluation

The evaluation of the proposed approach is reported on in this section. For the evaluation, as noted earlier, a U&E data set provided by Arrowe Park Hospital in Merseyside in the UK was used. The data set was entirely anonymised and ethical approval for its usage in anonymised form obtained by Arrowe Park Hospital. Details concerning this data set are given in Subsect. 5.1 below. The Objectives of the evaluation were

1. To investigate the most appropriate imputation strategy for addressing the unequal length of the pathology time series to be considered.
2. To determine the overall performance of the proposed approach using a proxy ground truth, as proposed in [7], and comparing with previously proposed approaches.

The first is considered in Subsect. 5.2 and the second in Subsect. 5.3. The value for χ was determined using the classifications in the training data. Five-cross validation was used through out. The average value for χ was found to be 0.57. All the experiments were run using a windows 10 desktop machine with a 3.2 GHz Quad-Core IntelCore i5 processor and 24 GB of RAM. For the LSTM, a GPU was used fitted with a NVIDA GeForceRTX 2060 unit.

5.1 Evaluation Data Set

The evaluation data set used was provided by Arrowe Park Hospital in Mersey-side in the UK. A general format of the data was presented in Sect. 3. The data set comprised 3,734 patient records with five U&E task results (time series) per patient. The operation of prediction models is typically conducted using a test set that features known values for the variable to be predicted which can be compared with the predicted values produced by the model. However, as noted in the introduction to this paper such test data is typically not available because of the resource required. Indeed this was the motivation for the work presented in this paper. As also noted earlier, in [7] a proxy ground truth was used. The same approach was therefore adopted with respect to the evaluation of the proposed approach. The final destinations of the patients within the U&E data set were used to create a proxy ground truth; whether they ended-up as emergency, in or out patients; equating to high, medium and low priority respectively ($C = \{high, medium, low\}$). The proxy ground truth data set comprised 255 patients with high priority, 123 with medium priority and 3,356 with low priority, covering all five tasks.

5.2 Data Imputation

As noted earlier, the interval between pathology results (points in the multi-variate time series), and the overall length of the pathology multi-variate time series, was variable. This is illustrated, using the U&E data set, in Figs. 3 and 4. Figure 3 shows the number of days between the individual patients considered. For the figure the patients were ordered according to the maximal interval in their pathology history and each given a sequential ID number. In Fig. 3, sequential ID numbers are listed on the x-axis, and maximal intervals on the y-axis. From the figure it can be seen that the majority of patients have a maximum pathology interval of less than 100 days. Figure 4 shows the range of time series lengths. The figure was generated by grouping the time series lengths, in the U&E evaluation data, into 10 bins of 6 starting with length 1, this covering time series from 1 to 60. From, the figure it can be seen that the majority of time series fell into the first bin, lengths 3 to 6.

LSTM model generation requires all time series to be of the same length. We would also like our time series to reflect the correct spacing between pathology results (we could have simply assumed a unit spacing). To engineer this we experimented with five alternative strategies:

1. **Mean imputation:** Using the mean value in the time series to impute additional values.

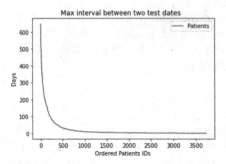

Fig. 3. Maximal interval in days per patient

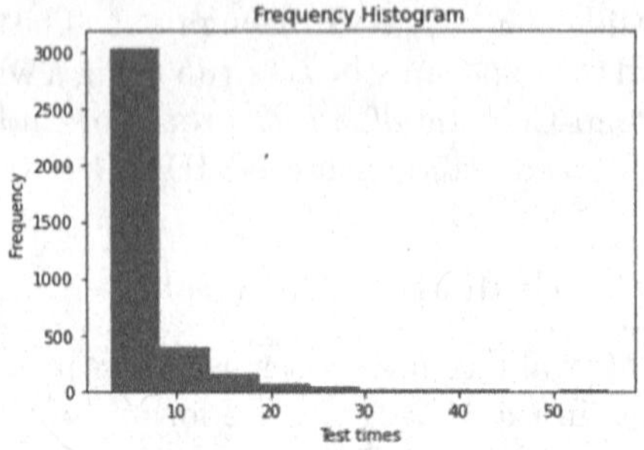

Fig. 4. Frequency of occurrence of different time series lengths

2. **Median imputation:** Using the medium value in the time series to impute additional values.
3. **Mode imputation:** Using the mode value in the time series to impute additional values.
4. **Zero imputation:** Using the value 0 to impute additional values.
5. **Padding and masking:** Skip missing time series values.

To evaluate these different strategies 'loss plots" were generated for each strategy as shown in Fig. 5. The number of epochs is given on the x-axis, and the loss in terms of Mean Squared Error (MSE) on the y-axis (note the y-axis scale is different for each graph). The same number of epochs was used in each case. The plots show the "training history" of the LSTMs. Each plot shows the loss between the training and test (validation) data as the model generation progressed. We want the loss to be minimal once model generation is complete. From the figure it can be seen that in all cases the training and test curves converge. Closer inspection indicates that the "padding and masking" method achieved the minimum loss.

Table 1 gives the classification accuracy, precision and recall values obtained, using the five imputation strategies when the proposed bounding box classification was applied (best results in bold font). The reported results are averages obtained using five-cross validation. From the table it can be seen that best performance was obtained using padding and masking (confirming the results from Fig. 5). As noted in Sect. 4, Padding and Masking is considered to offer the advantage that it preserves the original length and irregular spacing of the time series; this seems to be the most appropriate method for dealing with the differing lengths of time series as it does not change the original information provided by the data itself. Zero imputation produced the worst performance.

5.3 Overall Performance

Table 2 gives the best results from Table 1 compared to the results reported in [7] and [8] where a very similar proxy ground truth was used; best results are again highlighted in bold font. In [7] two classification models were considered a

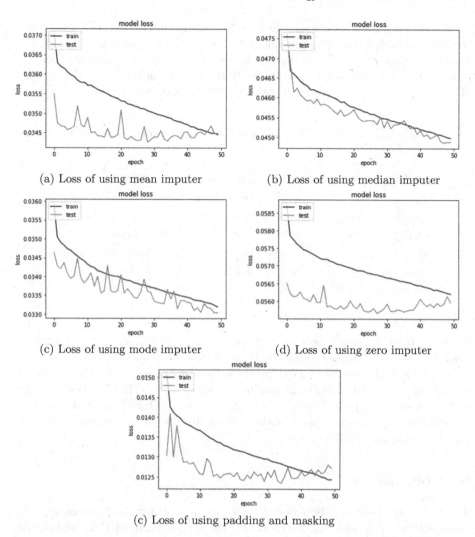

(a) Loss of using mean imputer

(b) Loss of using median imputer

(c) Loss of using mode imputer

(d) Loss of using zero imputer

(c) Loss of using padding and masking

Fig. 5. Performance comparison of the five alternative imputation strategies in terms of LSTM model generation

Table 1. Performance comparison of the five alternative imputation strategies in terms of accuracy, precision and recall

Imputation strategy	Acc.	Precision			Recall		
		High	Med.	Low	High	Med.	Low
Mean	0.56	0.60	0.51	0.53	**0.59**	0.59	**0.75**
Median	0.33	0.57	0.29	0.16	0.31	0.38	0.29
Mode	0.60	0.60	0.51	0.41	0.55	0.57	0.50
Zero	0.21	0.28	0.21	0.17	0.27	0.24	0.31
Pad. & Mask.	**0.73**	**0.69**	**0.62**	**0.61**	0.58	**0.69**	0.47
Ave	0.48	0.55	0.43	0.38	0.43	0.49	0.46

k Nearest Neighbour model and an LSTM model, the best results for both are included in Table 2. In a [8] an Anomaly Detection (AD) approach was proposed. Two versions were considered, a point-based AD approach and a time series AD approach. The best reported results for both are also included in Table 2. Note that in [8] only average precision and recall were reported.

Table 2. Average Accuracy, Precision and Recall results compared with the results reported in [7] and [8]

Method	Acc.	Precision			Recall		
		High	Med.	Low	High	Med.	Low
LSTM and bounding box	**0.73**	**0.69**	**0.62**	0.61	0.58	**0.69**	0.47
LSTM [7]	0.61	0.58	0.55	**0.69**	**0.79**	0.59	0.63
kNN [7]	0.60	0.42	0.51	0.85	0.70	0.55	**0.75**
Point-based AD [8]	0.34	0.35			0.43		
Time series AD [8]	0.45	0.45			0.43		

From Table 2 it can be seen that the proposed LSTM prediction coupled with bounding box classification produced the best overall accuracy. Closer inspection of the table indicates that good precision was obtained, using the proposed method, with respect to high and medium priority classes. The poor performance of the anomaly detection approaches (Point-based AD and Time Series AD) is probably because anomalous pathology results do not necessarily equate to priority pathology results.

6   Conclusions

The work presented in this paper was directed at multi-variate pathology data prioritisation in the absence of a ground truth. This is typically the case because of the resource required. A proposed approach has been presented that used LSTM prediction couple with a novel bounding box classification mechanism. The main contributions of this paper are: (i) the idea of using LSTM predicted test results as a marker for prioritisation and (ii) the derived "bounding box" technique for prioritisation classification. A further challenge was the irregular sampling interval of pathology data, and the variable length of the pathology history associated with a given pathology result. A number of alternative imputation techniques were therefore considered. The operation of the proposed approach was compared with alternative techniques from the literature using a proxy ground truth. The results indicated a better overall performance than that achieved by earlier work. A best accuracy of 73% was obtained. Padding and masking was found to be the most appropriate method for ensuring all time series were of the same size. For future work the authors intend to investigate:

(i) the generation of artificial evaluation data sets to provide for a more comprehensive evaluation, and (ii) a comprehensive collaborate with clinicians to obtain feed back regarding the prioritisation produced and to test the utility of the best performing mechanism in a real setting. The authors are currently liaising with domain experts on the practical impact of the proposed pathology data prioritisation mechanism presented this paper.

References

1. Ahmed, R., Nasiri, F., Zayed, T.: A novel neutrosophic-based machine learning approach for maintenance prioritization in healthcare facilities. J. Build. Eng. **42**, 102480 (2021)
2. Chandralekha, M., Shenbagavadivu, N.: Data analytics for risk of hospitalization of cardiac patients. IETE J. Res. 1–10 (2021)
3. Doyle, R., et al.: Machine learning-based prediction of COVID-19 mortality with limited attributes to expedite patient prognosis and triage: retrospective observational study. JMIRx Med **2**(4), e29392 (2021)
4. Jiang, H., et al.: Machine learning-based models to support decision-making in emergency department triage for patients with suspected cardiovascular disease. Int. J. Med. Informat. **145**, 104326 (2021)
5. Klang, E., et al.: Predicting adult neuroscience intensive care unit admission from emergency department triage using a retrospective, tabular-free text machine learning approach. Sci. Rep. **11**(1), 1–9 (2021)
6. Ochella, S., Shafiee, M., Sansom, C.: Adopting machine learning and condition monitoring P-F curves in determining and prioritizing high-value assets for life extension. Expert Syst. Appl. **176**, 114897 (2021)
7. Qi, J., Burnside, G., Charnley, P., Coenen, F.: Event-based pathology data prioritisation: a study using multi-variate time series classification. In: Proceedings of the 13th International Joint Conference on Knowledge Discovery, Knowledge Engineering and Knowledge Management - KDIR, pp. 121–128. INSTICC, SciTePress (2021)
8. Qi, J., Burnside, G., Coenen, F.: Ranking pathology data in the absence of a ground truth. In: Bramer, M., Ellis, R. (eds.) SGAI-AI 2021. LNCS (LNAI), vol. 13101, pp. 209–223. Springer, Cham (2021). https://doi.org/10.1007/978-3-030-91100-3_18
9. Sak, H., Senior, A., Beaufays, F.: Long short-term memory based recurrent neural network architectures for large vocabulary speech recognition (2014). arXiv preprint arXiv:1402.1128
10. Sobrinho, Á., Dias, L., da Silva, E., Candeia, A.P., et al.: Machine learning classification models for COVID-19 test prioritization in Brazil. J. Med. Internet Res. **23**(4), e27293 (2021)
11. Weerakody, P.B., Wong, K.W., Wang, G., Ela, W.: A review of irregular time series data handling with gated recurrent neural networks. Neurocomputing **441**, 161–178 (2021)
12. Wilkin, C., Ferreira, A., Rotaru, K., Gaerlan, L.R.: Big data prioritization in SCM decision-making: its role and performance implications. Int. J. Account. Inf. Syst. **38**, 100470 (2020)
13. Wu, J., Li, N., Zhao, Y.: Missing data filling based on the spectral analysis and the long short-term memory network. In: 2021 International Symposium on Computer Technology and Information Science (ISCTIS), pp. 198–202. IEEE (2021)

14. Xie, F., et al.: Score for emergency risk prediction (SERP): an interpretable machine learning autoscore-derived triage tool for predicting mortality after emergency admissions. medRxiv (2021)
15. Zaman, M.A.U., Dongping, D.: A stochastic multivariate irregularly sampled time series imputation method for electronic health records. BioMedInformatics **1**(3), 166–181 (2021)

Outlier/Anomaly Detection of Univariate Time Series: A Dataset Collection and Benchmark

David Muhr[1,2]([✉])[iD] and Michael Affenzeller[2,3][iD]

[1] BMW Group, Steyr, Austria
david.muhr@bmw.com
[2] Johannes Kepler University, Linz, Austria
[3] University of Applied Sciences Upper Austria, Hagenberg, Austria

Abstract. In this paper, we present an extensive collection of outlier/anomaly detection tasks to identify unusual series from a given time series dataset. The presented work is based on the popular UCR time series classification archive. In addition to the detection tasks, we provide curated benchmarks, an evaluation scheme and baseline results. The resulting unusual time series detection collection is openly available at: https://outlier-detection.github.io/utsd/.

Keywords: Outlier detection · Anomaly detection · Time series

1 Introduction

Frustrated by the difficulty of evaluating and comparing rival time series classification approaches, Keogh and Folias introduced the University of California Riverside (UCR) time series archive in 2002 [10]. The archive is continually being expanded and currently contains 128 datasets [6]. Besides time series classification, another important time series data mining task is *time series outlier detection* also referred to as *time series anomaly detection* [8]. We use the UCR time series archive as a basis to define a range of outlier detection tasks and benchmarks. An outlier or anomaly is frequently defined as "an observation (or subset of observations) which appears to be inconsistent with the remainder of that set of data" [1]. Time series outlier detection commonly refers to three separate tasks: (1) the detection of *outlier points* in time series, (2) the detection of *outlier subsequences* in time series, and (3) the detection of *outlier series*, which is the focus of our work. Hyndman et al. [9] refer to the detection of entire outlier series as *unusual time series detection*, to differentiate the task from point and subsequence outlier detection, which we adopt in the following sections.

2 Related Work

The evaluation of outlier detection algorithms has been identified as a constant challenge in outlier detection research [4]. Lai et al. [11] recently introduced an

© The Author(s), under exclusive license to Springer Nature Switzerland AG 2022
R. Wrembel et al. (Eds.): DaWaK 2022, LNCS 13428, pp. 163–169, 2022.
https://doi.org/10.1007/978-3-031-12670-3_14

extensive dataset collection for point and subsequence outliers in time series, which focuses on synthetic data generation with a small number of real-world datasets. The detection of unusual time series is another frequently described task. To the best of our knowledge, there exists no standardized dataset collection to evaluate outlier detection methods for such a task. Previous authors working on this kind of task use synthetic or proprietary datasets [5,9,12,16], or (a small selection of) repurposed time series classification datasets [2,3,13,15].

3   Dataset Collection

We further refer to the proposed collection of tasks and benchmarks as the unusual time series detection collection (UTSD). Recall that the UCR time series archive is continually being expanded, and we thus version the tasks and benchmarks to account for future updates. We define the first version of the UTSD collection as **UTSDv1**. Emmott et al. [7] define four requirements for the systematic construction of outlier detection benchmark datasets:

Requirement 1. Normal points should be drawn from a real-world process.
Requirement 2. Anomaly points should be drawn from a real-world process.
Requirement 3. Many benchmark datasets are needed.
Requirement 4. Benchmark datasets' difficulties should be characterized.

To address the outlined requirements, we prepare curated benchmarks that contain only a subset of the tasks in the collection, for example, by excluding synthetic datasets, to fulfill requirements 1 and 2. Requirement 3 can be satisfied by choosing a large enough selection of UCR datasets as a basis for the proposed benchmarks. The UCR archive provides an overview of classification results, which can be used as a measure of difficulty to fulfill requirement 4.

3.1   Detection Tasks

For each percentage $p \in \{2.5\%, 5\%, 10\%\}$ of outliers we define the following detection tasks. We rank the classes of each dataset according to their class count and class name order, which we define as the class normality, i.e., classes with higher class counts are considered 'normal' and, if classes contain an equal amount of elements, the lower class according to name is considered 'normal'. Note that integer values describe the class names in the UCR archive, allowing us to define the class name order based on its value. For all outlier detection tasks, the outlier class is defined as a random sample of all non-normal classes, such that the final dataset contains at most p percent of outliers. As proposed by [4], we attempt to mitigate the impact of randomization when downsampling by repeatedly sampling each dataset ten times, resulting in ten different variants for each task. Note that two datasets are treated differently based on their semantic interpretation, namely the binary *Strawberry* and *ItalyPowerDemand* datasets. The Strawberry dataset encompasses the minority class 'Strawberry' (authentic samples) and the majority class 'Non-Strawberry' (adulterated strawberries and other fruits), which we define as outliers. The ItalyPowerDemand dataset differentiates between the class 'October to March' (549 samples) and the class 'April to September' (547 samples). Because of the summer holidays in August, the majority class shows outlier behavior; hence we choose it as the outlier class.

Single-Concept Normality. We define the single-concept normality task as follows: The normal class is defined as the first majority class. There are cases in which a percentage p would lead to a dataset with zero outliers; in this case, the task is not available for the specific dataset and percentage. None of the datasets initially contains less than 10% of outliers.

Multi-concept Normality. We define the multi-concept normality tasks as follows: The normal classes are defined as the c first majority classes with $c > 1$, which we compare to the base case $c = 1$. Therefore, multi-concept tasks are only defined for datasets with at least three classes. The UCR time series archive currently contains 86 such multiclass datasets.

3.2   Detection Benchmarks

The detection benchmarks are a curated selection of the proposed tasks, making it easier to compare novel outlier detection methods to previously published work. We provide a single-concept benchmark as well as a multi-concept benchmark. For both benchmarks, we define the following conditions:

1. The dataset must not contain missing values.
2. The dataset must not be created synthetically.
3. The dataset must not be a duplicate of another benchmark dataset.
4. The dataset must not contain less than two outliers.

Points 1 and 2 follow from the requirements proposed in [7]. Point 3 refers to the fact that the UCR time series archive contains datasets that are very closely related to each other, e.g., downsampled variants. The last point ensures that at least one sample per class is contained in a stratified evaluation setting. We drop all datasets that violate these conditions and do not prescribe additional data preparation steps, e.g., for normalization or handling of missing values, other than the steps already performed in the UCR time series classification archive.

UTSDv1 Single-Concept Benchmark. We require that the normal class contains at least 300 data points for the single-concept benchmark, which results in 39 possible datasets. Four of the datasets[1] are simulated, ten of the datasets are considered duplicates[2,3,4,5,6], and one dataset contains missing values[7], which results in 24 datasets for the single-concept benchmark as visible in Table 1.

---

[1] *CBF, ChlorineConcentration, Mallat* and *TwoPatterns.*

[2] *FreezerSmallTrain* is a downsampled version of *FreezerRegularTrain.*

[3] *FacesUCR* is considered a duplicate of *FacesAll.*

[4] *MixedShapesSmallTrain* is a downsampled version of *MixedShapesRegularTrain.*

[5] *DistalPhalanxOutlineAgeGroup, DistalPhalanxOutlineCorrect, MiddlePhalanxOutlineCorrect,* and *ProximalPhalanxOutlineCorrect* are considered duplicates of the *PhalangesOutlinesCorrect.*

[6] *UWaveGestureLibraryX/Y/Z* are considered duplicates of *UWaveGestureLibraryAll.*

[7] *MelbournePedestrian.*

UTSDv1 Multi-concept Benchmark. In the multi-concept benchmark, we compare $c = 1$ to $c = 2$ and $c = 3$. We only consider datasets with at least 100 data points in the first majority class, which results in 300 points with $c = 3$ if the classes are distributed equally, resulting in 82 possible datasets. Furthermore, we only consider datasets that contain at least two outliers for all percentages and all numbers of concepts, leaving 34 possible datasets. Out of the 34 remaining datasets, five of them contain missing values[8], three of them are simulated[9] and seven are considered duplicates[10,11]. This results in 19 datasets for the benchmark as visible in Table 2.

Table 1. UTSDv1 single-concept benchmarks for different percentages p.

Dataset	n-total	n-normal	$p = 2.5\%$	$p = 5.0\%$	$p = 10.0\%$
CinCECGTorso	1420	355	9	18	39
Crop	24000	1000	25	52	111
ECG5000	5000	2919	74	153	324
ECGFiveDays	884	442	11	23	49
Earthquakes	461	368	9	19	40
ElectricDevices	16637	4275	109	225	475
FaceAll	2250	327	8	17	36
FordA	4921	2527	64	133	280
FordB	4446	2261	57	119	251
FreezerRegularTrain	3000	1500	38	78	166
HandOutlines	1370	875	22	46	97
ItalyPowerDemand	1096	547	14	28	60
MedicalImages	1141	594	15	31	66
MixedShapesRegularTrain	2925	754	19	39	83
MoteStrain	1272	685	17	36	76
PhalangesOutlinesCorrect	2658	1698	43	89	188
SemgHandGenderCh2	900	540	13	28	60
SonyAIBORobotSurface2	980	604	15	31	67
StarLightCurves	9236	5327	136	280	591
Strawberry	983	351	9	18	39
TwoLeadECG	1162	581	14	30	64
UWaveGestureLibraryAll	4478	560	14	29	62
Wafer	7164	6402	164	336	711
Yoga	3300	1770	45	93	196

3.3   Evaluation and Baselines

The most popular evaluation measure in unsupervised outlier detection is based on the Receiver Operating Characteristic (ROC). A ROC can be summarized

---

[8] *AllGestureWiimoteX/Y/Z, MelbournePedestrian* and *PLAID*.

[9] *Mallat, SyntheticControl* and *TwoPatterns*.

[10] *DistalPhalanxTW* and *MiddlePhalanxTW* are considered duplicates of *ProximalPhalanxTW*, which contains the largest sample count of the three datasets.

[11] See Footnotes 3, 4 and 6.

Table 2. UTSDv1 multi-concept benchmarks for percentages p and concepts c.

Dataset	n-total	n-normal	$p = 2.5\%$	$p = 5.0\%$	$p = 10.00\%$
CinCECGTorso	1420	355/710/1065	9/18/27	18/37/56	39/78/118
Crop	24000	1000/2000/3000	25/51/76	52/105/157	111/222/333
ElectricDevices	16637	4275/8462/11101	109/216/284	225/445/584	475/940/1233
EthanolLevel	1004	252/504/754	6/12/19	13/26/39	28/56/83
FaceAll	2250	327/600/787	8/15/20	17/31/41	36/66/87
FiftyWords	905	109/200/261	2/5/6	5/10/13	12/22/29
Haptics	463	100/200/293	2/5/7	5/10/15	11/22/32
InlineSkate	650	117/225/328	3/5/8	6/11/17	13/25/36
InsectWingbeatSound	2200	200/400/600	5/10/15	10/21/31	22/44/66
MedicalImages	1141	594/706/812	15/18/20	31/37/42	66/78/90
MixedShapesRegularTrain	2925	754/1435/1993	19/36/51	39/75/104	83/159/221
Phoneme	2110	238/387/532	6/9/13	12/20/28	26/43/59
ProximalPhalanxTW	605	252/428/526	6/10/13	13/22/27	28/47/58
SemgHandMovementCh2	900	150/300/450	3/7/11	7/15/23	16/33/50
SemgHandSubjectCh2	900	180/360/540	4/9/13	9/18/28	20/40/60
Symbols	1020	181/362/529	4/9/13	9/19/27	20/40/58
UWaveGestureLibraryAll	4478	560/1120/1680	14/28/43	29/58/88	62/124/186
WordSynonyms	905	200/315/387	5/8/9	10/16/20	22/35/43
Worms	258	109/154/198	2/3/5	5/8/10	12/17/22

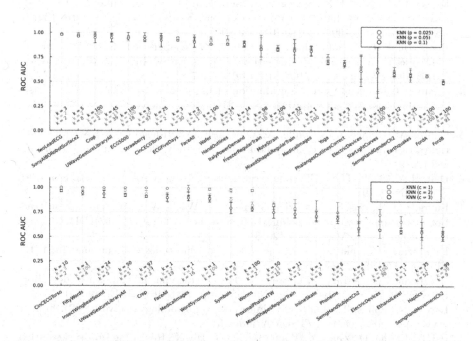

Fig. 1. Baselines for the UTSDv1 single-concept benchmark for $p \in \{2.5\%, 5\%, 10\%\}$ and the UTSDv1 multi-concept benchmark for $c \in \{1, 2, 3\}$ and $p = 5\%$.

by a single value known as the area under the ROC curve (ROC AUC) [4]. We provide a k-nearest neighbor (k-NN) baseline detector as described in [14]. We do not employ additional data preprocessing but optimize the number of neighbors for all $k \in \{1, 2, \ldots, 100\}$. We report the mean ROC AUC score and corresponding standard deviation over the ten randomly-sampled variants and the best k for each dataset. Each score is the result of a repeated, two-fold, stratified cross-validation, which we suggest as the default evaluation scheme. Figure 1 shows the results for both benchmarks and the best value of k for a task, which, in the case of draws, represents the higher value.

4   Conclusions

In this work, we present the first standardized collection of tasks and benchmarks for the detection of unusual time series. The proposed UTSD collection will aid future researchers in developing and comparing novel time series outlier detection methods. To facilitate reproducibility, we publish all data and code at https://github.com/outlier-detection/utsd/.

References

1. Barnett, V., Lewis, T.: Outliers in Statistical Data. Wiley, New York (1978)
2. Beggel, L., Kausler, B.X., Schiegg, M., Pfeiffer, M., Bischl, B.: Time series anomaly detection based on shapelet learning. Comput. Statistics **34**(3), 945–976 (2018). https://doi.org/10.1007/s00180-018-0824-9
3. Benkabou, S.-E., Benabdeslem, K., Canitia, B.: Unsupervised outlier detection for time series by entropy and dynamic time warping. Knowl. Inf. Syst. **54**(2), 463–486 (2017). https://doi.org/10.1007/s10115-017-1067-8
4. Campos, G.O., et al.: On the evaluation of unsupervised outlier detection: measures, datasets, and an empirical study. Data Min. Knowl. Disc. **30**(4), 891–927 (2016)
5. Canizo, M., Triguero, I., Conde, A., Onieva, E.: Multi-head CNN-RNN for multi-time series anomaly detection: an industrial case study. Neurocomputing **363**, 246–260 (2019)
6. Dau, H.A., et al.: The UCR time series archive. IEEE/CAA J. Autom. Sin. **6**(6), 1293–1305 (2019)
7. Emmott, A.F., Das, S., Dietterich, T., Fern, A., Wong, W.K.: Systematic construction of anomaly detection benchmarks from real data. In: ACM SIGKDD, Chicago, Illinois, pp. 16–21. ACM Press (2013)
8. Fakhrazari, A., Vakilzadian, H.: A survey on time series data mining. In: IEEE EIT, pp. 476–481 (2017)
9. Hyndman, R.J., Wang, E., Laptev, N.: Large-scale unusual time series detection. In: IEEE ICDMW, Atlantic City, NJ, USA, pp. 1616–1619. IEEE (2015)
10. Keogh, E., Folias, T.: The UCR time series data mining archive (2002). http://www.cs.ucr.edu/~eamonn/TSDMA/index.html
11. Lai, K.H., Zha, D., Zhao, Y., Wang, G., Xu, J., Hu, X.: Revisiting time series outlier detection: definitions and benchmarks. In: NeurIPS. Datasets and Benchmarks Track, p. 13 (2021)

12. Laptev, N., Amizadeh, S., Flint, I.: Generic and scalable framework for automated time-series anomaly detection. In: Proceedings of the 21th ACM SIGKDD International Conference on Knowledge Discovery and Data Mining, Sydney, NSW, Australia, pp. 1939–1947. ACM (2015)
13. Lara, J.A., Lizcano, D., Rampérez, V., Soriano, J.: A method for outlier detection based on cluster analysis and visual expert criteria. Expert Syst. **37**(5), e12473 (2020)
14. Ramaswamy, S., Rastogi, R., Shim, K.: Efficient algorithms for mining outliers from large data sets. SIGMOD Rec. **29**(2), 427–438 (2000)
15. Rebbapragada, U., Protopapas, P., Brodley, C.E., Alcock, C.: Finding anomalous periodic time series: an application to catalogs of periodic variable stars. Mach. Learn. **74**(3), 281–313 (2009)
16. Sodja, C.: Detecting anomalous time series by GAMLSS-Akaike-Weights-Scoring. J. Comput. Graph. Stat. 1–13 (2021)

Schema Discovery and Construction

Automatic Machine Learning-Based OLAP Measure Detection for Tabular Data

Yuzhao Yang[1(✉)], Fatma Abdelhédi[3], Jérôme Darmont[2], Franck Ravat[1], and Olivier Teste[1]

[1] IRIT-CNRS (UMR 5505), Université de Toulouse, Toulouse, France
{Yuzhao.Yang,Franck.Ravat,Olivier.Teste}@irit.fr
[2] Univ Lyon, Univ Lyon 2, UR ERIC, Lyon, France
jerome.darmont@univ-lyon2.fr
[3] CBI² - TRIMANE, Paris, France
Fatma.Abdelhedi@trimane.fr

Abstract. Nowadays, it is difficult for companies and organisations without Business Intelligence (BI) experts to carry out data analyses. Existing automatic data warehouse design methods cannot treat with tabular data commonly defined without schema. Dimensions and hierarchies can still be deduced by detecting functional dependencies, but the detection of measures remains a challenge. To solve this issue, we propose a machine learning-based method to detect measures by defining three categories of features for numerical columns. The method is tested on real-world datasets and with various machine learning algorithms, concluding that random forest performs best for measure detection.

Keywords: Data warehouses · OLAP · Measure detection · Tabular data

1 Introduction

Business Intelligence (BI) plays an important role in numerous companies and organizations to efficiently support decision making processes. In classical BI architectures, data from heterogeneous sources are integrated into a Data Warehouse (DW) usually modeled in a multidimensional way, allowing decision makers to analyze data by On-Line Analytical Processing (OLAP) [11]. A multidimensional DW organizes data according to analysis subjects (facts) associated with analysis axes (dimensions). Dimension attributes may be ordered according to their granularity (hierarchies) and each fact is composed of indicators (measures).

With the development of information systems and the availability of numerous open datasets, various data become much more accessible to enterprises, organizations and even individuals, who have data analysis needs to help them take decisions [1]. However, the design of a DW is typically carried out manually, requires expert knowledge and BI experience [26], may be time-consuming and costly. Thence, automating the DW design process is desirable to allow businesses and organizations take advantage of BI.

There are different automatic or semi-automatic approaches for the design of multidimensional DW schemas [25]. However most of these methods focus on data sources with explicit schema: relational data with Entity-Relationship (ER) schema, XML data with Document Type Definitions (DTDs), etc. Nevertheless, tabular data such as spreadsheet data and Comma Separated Value (CSV) files are very common in enterprises, and even more in the open data world. They may also be DW data sources, but whose schema is not available. Building dimensions and hierarchies for tabular data can be done by detecting functional dependencies [32]. In the existing methods for other sources, measures are defined manually by users or are detected with respect to data types (numerical values) and cardinalities, which is impractical for tabular data without schema. Yet, measures remain central elements in multidimensional models, as they are the indicators that assess the analyzed activities. **Therefore, measure detection for automatic DW design from tabular data is an important task**. To the best of our knowledge, there is no specific approach addressing this challenge.

Tabular data may bear quite simple or very complex structures [2]. Simple structures consist of one header row followed by rows containing data values. Headers label the data rows below, while data rows contain tuples akin to relational database tuples. Most CSV files bear a simple structure, while spreadsheet files and HTML tables can be more complex, e.g., cross tables [17]. Such tables contain two or several dimensions, and may also contain several dimension levels. Moreover, there also exists other complex structures such as concise tables, nested tables, multivalued tables and split tables [17].

The data region can be extracted from a cross tables by some algorithms [5,6,16,31] where measures are located. The other types of complex structures can be converted into simple structures [6]. However, for simple-structured tabular data, DW elements cannot be directly extracted either without a schema or metadata, as the data do not bear a particular layout. Measures are usually numerical data, but numerical columns are not necessarily measures, since there also exists descriptive numerical attributes. Thus, we intend to find the numerical columns that conform to the characteristics of measures. We hypothesize that there are differences in terms of features between numerical data that are potential measures and those that are not. **Therefore, in this paper, we define specific features for numerical columns and propose a machine learning-based method to automatically detect measures.**

The remainder of this paper is organized as follows. In Sect. 2, we review the related works about measure selection for automatic DW design. In Sect. 3, we detail and discuss the measure detection process and the features we propose. In Sect. 4, we present and interpret our experimental results. Finally, in Sect. 5, we conclude this paper and hint at future research.

2   Related Works

There are various methods dedicated to automatic or semi-automatic DW schema design, with different measure selection approaches. Since the selected measures should correspond to business requirements, in many methods [8,9,14, 27], they are assigned directly by the users.

In a semi-automatic method to model DW from E/R diagrams [13], the fact table is selected by calculating the Connection Topology Value (CTV) of each entity, which is a composite function of the topology value of direct and indirect one-to-many relationships. Measures are still chosen manually by the user, but the scope of the choice is reduced. Moreover, another approach fully automates DW design from an E/R schema [22]. In contrast, some approaches aim at discovering measures, including 1) selecting many-to-many relationships containing numeric and additive non-key facts [15]; 2) analyzing business queries for data items indicating business performance [4]; 3) basing on most frequently updated entities [10]; or 4) selecting the numerical data that can be aggregated [30]. All these approaches work in the context of automatic DW design based on E/R schemas. However, the constraints and relationships mentioned in such methods cannot be directly applied on tabular data.

Another trend is using knowledge-based methods for automatic DW design [28]. Key information on measures and dimensions are extracted through a Natural Language Processing (NLP) model based on sentences from the business requirements. Candidate measures are all numerical column and are then validated by some constraints in a predefined domain ontology and by checking whether they can be aggregated. However, the method needs business requirements to train the NLP model. Defining the domain ontology is also difficult.

In summary, there is no specific method to automatically detect measures from tabular data in the absence of schema and explicit business requirements. Thus, we propose a machine learning-based method for measure detection from tabular data.

3   Measure Detection

3.1   Overview

Figure 1 shows an overview of our measure detection process for tabular data. If tabular data bear a complex structure, we use table structure detection algorithms [5,6,16,31] to verify whether data lie in a cross table. If so, measures are extracted from the data region, save aggregated values are excluded. Otherwise, data are converted by the algorithm proposed in [6] into a simple structure that is formally defined in Definition 2.

Definition 1. *Measures are numerical and quantitative attributes of the analysis subject evaluating the activities of an organisation and that can be aggregated with respect to dimensions. They can be additive, semi-additive or non-additive [12].*

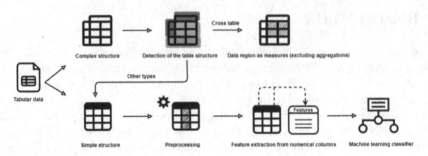

Fig. 1. Measure detection for tabular data

Definition 2. *A tabular dataset of simple structure TS is defined as $\{C, R, A, V\}$, where:*

- $C = \{C_1, C_2, ...C_{n_c}\}$ *is a set of columns, where n_c is the number of columns in TS. For a given column $C_i \in C$, index i corresponds to the column's position in TS. The number of non-null values in column C_i is denoted as $n_t(C_i)$. The number of non-null distinct values is denoted as $n_u(C_i)$;*
- $R = \{R_1, R_2, ..., R_{n_r}\}$ *is a set of rows (excluding the first, header row), where n_r is the number of non-header rows in TS. For a given row $R_j \in R$, j represents the index of the row corresponding to its position in TS;*
- $A = \{A_{C_1}, A_{C_2}, ..., A_{C_{n_c}}\}$ *is a set of attribute headers. For a given attribute header $A_{C_i} \in A$, C_i represents the column labeled by A_{C_i};*
- V *is a matrix of cell values whose dimension is $n_r \times n_c$. For a given cell value $V_{R_j,C_i} \in V$, R_j and C_i are the row and the column where the cell is located, respectively.*

In the following sections, we focus on measure detection for tabular data of simple structure. Since measures are numerical, we regard all numerical columns as candidates. Yet, preprocessing the dataset is necessary for the selection of numerical columns. Then, to distinguish between measure and non-measure numerical columns we extract features from numerical columns and use machine learning classifiers to estimate whether they are measures.

3.2   Preprocessing

As candidate measures are numerical columns, we must first identify numerical columns. If all values of a column are numerical, we easily identify numerical columns. However, there are sometimes columns containing numerical values with their unit, or columns containing both numerical and textual values used for replacing empty cells. Such mixed values must lead to numerical columns and require preprocessing.

Columns containing values with a unit are identified by verifying whether each cell o bear the same structure, e.g., "text + number" or "number + text".

We also verify whether the text of each column is the same or if it is categorical by using the algorithm proposed by [3]. Then, we extract numerical values via regular expressions and tag the column as numerical. Eventually, numerical columns containing empty values replaced by some text, e.g., "n/a", "null" or "unknown", are treated as numerical, with textual values being removed.

3.3 Feature Extraction

After the preprocessing phase, we extract the numerical columns' features. When defining features, we analyze both general information and some statistical characteristics of numerical columns. Since tabular data of simple structure are usually relational and may exhibit specific column positional habits, we also consider column inter-relationships. Features are thus subdivided into three categories: general features, statistical features and inter-column features. For a given numerical column C_i, we define the following features.

General Features. These features reflect basic information on numerical columns. Such general features may help check whether a numerical column is likely to be quantitative and help evaluate business activities. General features follow.

– **Data type:** $type = \begin{cases} 1 \text{ if } type(C_i) = integer \\ 0 \text{ if } type(C_i) = float \end{cases}$, where $type(C_i)$ is C_i's data type.

Intuitively, float data are more likely to be quantitative and to allow evaluating activities. For example, temperature, salary and sales amount are float data can be considered as measures in most cases.

– **Positive/Negative/Zero value ratio:** $rpos = \dfrac{n_{pos}(C_i)}{n_t(C_i)}$, $rneg = \dfrac{n_{neg}(C_i)}{n_t(C_i)}$, $rzero = \dfrac{n_{zero}(C_i)}{n_t(C_i)}$, where $n_{pos}(C_i)$, $n_{neg}(C_i)$ and $n_{zero}(C_i)$ are the number of positive, negative and zero values in C_i, respectively, and $n_t(C_i)$ is the number of non-null values in C_i.

These features may help identifying both qualitative and quantitative columns. Qualitative data values, e.g., ID or zip code, are rarely negative or equal to zero. Thus, when there are many zero and negative values in a column, it is more likely to be a measure.

– **Unique value ratio:** $runique = \dfrac{n_u(C_i)}{n_t(C_i)}$.

The unique value ratio can reveal some typological information about a column. For example, in a descriptive dataset, IDs are always unique, so the unique value ratio is always 1. In a dataset containing fact table data, keys and descriptive data may be repetitive, but equal measures should be quite scarce.

- **Same digital number**:

$$sdn = \begin{cases} 1 \text{ if } \forall i \in [1, n_t(C_i) - 1], nd_{R_j, C_i} = nd_{R_{j+1}, C_i} \wedge type(C_i) = integer \\ 0 \text{ if } (\exists i \in [1, n_t(C_i) - 1], nd_{R_j, C_i} \neq nd_{R_{j+1}, C_i} \wedge type(C_i) = integer) \\ \quad \vee (type(C_i) = float) \end{cases}$$

where nd_{R_j, C_i} is the number of digits in cell value V_{R_j, C_i}, which is calculated as $nd_{R_j, C_i} = floor(log_{10}^{V_{R_j, C_i}}) + 1$.

This feature tells whether all the values of an integer column have the same number of digits. If it is the case, the column is likely to be a nominal number [3] representing the name or identifier of an element that cannot be a measure. For example, the French social security number always contains 15 digits.

Statistical Features. Since candidate columns are numerical, statistical features must be considered. They can indeed reflect the distribution of column values. Statistical features follow.

- **Average/Minimum/Maximum/Median/Upper quartile/Lower quartile values**: $avg = avg(C_i)$, $min = min(C_i)$, $max = max(C_i)$, $median = median(C_i)$, $upquar = upquar(C_i)$ and $lowquar = lowquar(C_i)$ represent the average, minimum, maximum, median, upper quartile and lower quartile of C_i, respectively.

We consider these basic statistical metrics as features. In some specific columns, their values always vary in a certain range. Using these features can thus be helpful for capturing such statistical behaviours.

- **Coefficient of variation**:

$$coevar = \begin{cases} standdev(C_i) \text{ if } avg(C_i) = 0 \\ \dfrac{standdev(C_i)}{avg(C_i)} \text{ if } avg(C_i) \neq 0 \end{cases}$$

where $standdev(C_i)$ is C_i's standard deviation.

The standard deviation can depict the amount of dispersion of a column values. Measures or descriptive attributes may have different degrees of dispersion, but by using the coefficient of variation, which is the ratio of the standard deviation by the average, we achieve a standardized degree of dispersion. For example, given two attributes "price of phone" and "temperature of city", the average price is much higher than that of temperature. A price variation of 10 is relatively much lower than that of tempera. Since the coefficient of variation is a ratio, when the average is equal to 0, it does not exist. Here, we define that when the average is 0, the feature is equal to the standard deviation of the column.

- **Range ratio**: $rrange = \dfrac{max - min}{n_u(C_i)}$.

The range ratio calculates the range of values with respect to the number of distinct values. It is useful to identify some ordinal data, even if they occur repetitively. For example, if we have student numbers ranging from 1000 to 2000 in a tabular dataset, but also courses and grades, a student number may occur many times while the range ratio is always 1 no matter the number of occurrences.

Inter-Column Features. Measures are aggregatable and are normally accompanied with attributes by which they are aggregated, as per the "group by" SQL clause. Typically, attributes linked to aggregations are located before measures in the source file. Therefore, we consider inter-column features that take inter-column relationships into account in the whole dataset.

- **Location ratio:** $rloc = \dfrac{i-1}{n_c - 1}$.

In many tables, the identifier and some other basic information usually lie at the beginning positions, while measures are usually in the latter positions. Thus, we also take column location into account. However, different datasets have different number of columns, so we must normalize the location feature as a ratio ranging between 0 and 1.

- **Numerical column ratio:** $rnum = \dfrac{n_{num}}{n_c}$, where n_{num} is the number of numerical columns in the whole dataset.

The ratio of numerical column number by total column number is a table feature. While there are tabular data that only contain descriptive information, others include numerical columns that may be measures.

- **Multiple functional dependencies:**
$$severalfds = \begin{cases} 1 \text{ if } \exists fd \in fdset, (fd.rhs = A_{C_i}) \wedge (size(fd.lhs) > 1) \\ 0 \text{ else} \end{cases}$$
where $fdset$ is the functional dependency set of the dataset, $fd.rhs$ is the right hand side attribute of functional dependency fd and $size(fd.lhs)$ is the number of attributes in the left hand side of fd.

In existing methods that exploit data sources with schemas, many-to-many relationships are usually employed for measure detection. In a DW, we usually analyze a fact with respect to different dimensions and measure values depend on dimensions' primary keys. Thus, we consider whether there is a functional dependency with A_{C_i} depending on several attributes as a feature.

- **Numerical neighbor**:

$$numn = \begin{cases} 1 & \text{if } (i = 1 \land type(C_{i+1}) \in num) \lor (i = n_c \land type(C_{i-1}) \in num) \\ & \lor (i \neq 1 \land i \neq n_c \land type(C_{i+1}) \in num \land type(C_{i-1}) \in num) \\ 0.5 & \text{if } (i \neq 1 \land i \neq n_c \land type(C_{i+1}) \in num \land type(C_{i-1}) \notin num) \\ & \lor (i \neq 1 \land i \neq n_c \land type(C_{i+1}) \notin num \land type(C_{i-1}) \in num) \\ 0 & \text{else} \end{cases}$$

where $num = \{integer, float\}$.

In a tabular dataset, the columns describing similar information are often clustered together. Measures are also likely to be located close together, meaning that there are numerical columns in neighboring positions. Thus, we define this feature to see if neighbors of a column are also numerical. If so, the column is likely to be a measure.

4   Experimental Validation and Discussion

4.1   Experimental Conditions

Our experiments are conducted on an Intel(R) Core(TM) i5-10210U 1.60 GHz CPU with a 16 GB RAM. We use 9 datasets in our experiments, from the governmental open data sites of France (**FR**), Canada (**CA**), UK (**UK**) and US (**US**), the French Development Agency (**AFD**), the New Zealand's official data agency (**NZ**), the American Center for Disease Control and Prevention (**CDC**), the World Bank (**WB**) and Kaggle (**KG**). Each dataset contains numerous tables with numerical columns on which features are extracted to feed the algorithms. Moreover, they are classified into five domains (Table 3) including Economy (**ECO**), Health (**HLT**), Government (**GOV**), Environment (**ENV**) and Society (**SOC**). Complete information about these datasets are provided in Appendix.

We apply the following widely used Machine Learning (ML) classification algorithms [29] (available in Python 3.7): 1) an SVM classifier with an RBF kernel (**SVM**), 2) a decision tree classifier based on the CART algorithm (**DT**), 3) a random forest classifier (**RF**) and 4) a k-nearest neighbors classifier (**KNN**). Deep learning models are not employed because they are more suitable for interpreting images, sounds and texts [18], while we analyze numerical columns.

We define the ground truth by analyzing each dataset context according to its website's description, header semantics and metadata. We also uphold the criteria from Definition 1. Thence, for each dataset, we compute all our proposed features (Sect. 3.3) for each numerical column, and label them to build training and test sets. Empty values in columns are ignored and not counted.

4.2   Baseline Methods

Numerical Typology-Based Method (TP). In a previous work, we proposed to select measures with respect to the type of numerical attributes [32]. Numerical data may be classified into nominal data, ordinal data, intervals and ratios [3]. Algorithms can detect the different numerical types [3]. We identified the columns of interval and ratio types as DW measures.

Functional Dependency-Based Method (FDB). As we already mentioned, in existing methods aimed at data with schemas, measures are selected in tables exhibiting many-to-many relationships; in other words, columns that are functionally dependent on dimension primary keys. With this idea in mind, we detect functional dependencies (FDs) in tabular data and select as measures the numerical columns that are functionally determined by several, other attributes. The FD detection algorithm that we use is HyFD [21]. HyFD indeed achieves the best performance against the seven most cited and important algorithms that are tested in [19].

We employ the Metanome Web-based toolbox [20], which is developed by the HyFD designers, to implement HyFD. Moreover, we use the Python library selenium[1] to feed input files in Metanom and get the FDs automatically. The extracted FDs are also used for generating the values of feature **severalfds**.

4.3   Experimental Results

Algorithm Effectiveness. We run the two baseline methods from Sect. 4.2 and train models with our proposed features by four ML algorithms (Sect. 4.1) on all datasets (Sect. 4.1). The ML algorithms are run by pycaret[2] AutoML Python library where the hyperparameters are tuned automatically. For the model generality and feauture importance experiments, we run ML algorithms from the sklearn[3] Python library.

We use three performance metrics: Recall (**R**), Precision (**P**) and F-Measure (**F**), as follows. Let N_{mm} and N_{mn} be the number of measures predicted as measures and non-measures, respectively; and N_{nm} and N_{nn} the number of non-measure predicted as non-measures and measures, respectively.

$$\text{Then,} \quad \mathbf{R} = \frac{N_{mm}}{N_{mm} + N_{mn}}, \quad \mathbf{P} = \frac{N_{mm}}{N_{mm} + N_{nm}} \quad \text{and} \quad \mathbf{F} = \frac{2 \times Precision \times Recall}{Precision + Recall}.$$

Table 1 shows the resulting values of **R**, **P** and **F** where the results of ML algorithms are obtained through a 10-fold cross validation by merging all datasets and randomly split them into 10 folds. The distribution of the cross validation results is depicted in Fig. 2.

We observe that **RF** exhibits the best F-measure (94.82%) and the result is not more dispersed than that of the other algorithms. Thus, **RF** shows the best performance on the measure detection problem. We also observe that **TP** and **FDB** do not have a good effectiveness when predicting measures, but **FDB** performs better than **TP**. **TP**'s bad performance is due to: 1) interval and ratio numerical columns are not all measures, e.g., longitude and latitude; 2) numerical typology detection algorithm are not flexible enough to cope with real-world data, because they are based on fixed rules. Regarding **FDB**, a numerical column

---

[1] https://selenium-python.readthedocs.io.

[2] https://pycaret.org/.

[3] https://scikit-learn.org.

Table 1. Global results

	TP	FDB	RF	SVM	DT	KNN
R(%)	80.05	75.43	**96.64**	94.77	94.08	90.16
P(%)	73.57	77.50	**90.89**	78.44	88.44	87.61
F(%)	76.67	76.45	**93.65**	85.76	91.12	88.78

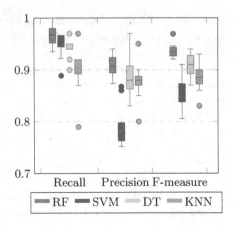

Fig. 2. Cross validation distribution

that is functionally determined by several other columns may not always be a measure, either. For example, let us consider a table describing sale facts with respect to customers and products, where sales' amount is indeed a measure. The customer ID is the customer dimension's primary key, but the customer's name and email may uniquely identify a customer, and thus may functionally determine the age of the customer, a numerical column that is not a measure.

Our ML-based measure detection method takes different types of features (Sect. 3.3) into account and can thus better handle the above exceptions and get better results.

Feature Category Effectiveness. To verify the effectiveness of each feature category we propose, we test different combinations of feature categories with our **RF**-based method. We first test single feature categories, combinations of two categories and then we compare the effectiveness of all categories. The result is shown in Table 2, where **GE** represents GEneral features, **ST** represents STatistical features and **IC** represents Inter-Column features. **ST** exhibits the best individual contribution. Yet, we can clearly see that combining feature categories achieves better performance in terms of recall, precision and F-measure, than using single feature categories. Ultimately, combining all feature categories yields the best performance. The results of applying other ML algorithms can be found in our github.

Table 2. Performance of feature categories and combinations with **RF**

	GE	ST	IC	GE+ST	GE+IC	ST+IC	ALL
R(%)	88.10	94.27	92.68	95.30	93.67	91.93	**96.64**
P(%)	83.59	86.28	80.91	88.21	86.13	91.14	**90.89**
F(%)	85.69	90.01	86.37	91.57	89.67	91.50	**93.65**

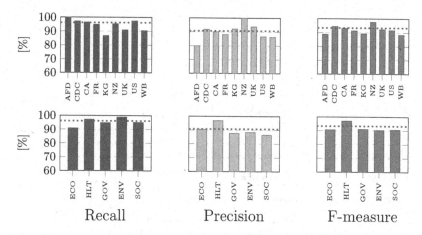

Fig. 3. Performance with respect to source and domain

Model Generality. To verify that the trained model achieved with our **RF**-based method is generic, we train data by excluding the datasets of one sources and test on them. We also carry out the same test by domain, i.e., economy (**ECO**), health (**HLT**), government (**GOV**), environment (**ENV**) and society (**SOC**). The results are shown in Fig. 3, where the charts above and below depict the results by source and domain, respectively. By comparing with former results, the difference of F-measure ranges from −5.02% to 4.23% for the test with respect to the source and from −3.17% to 3.36% for the test with respect to the domain. The trained model with the defined features is thus generic regardless of the source and the domain of data. The results of applying other ML algorithms can be found in our github[4].

Feature Importance. To analyze our different features, we compute the permutation importance (decrease in prediction accuracy when a feature is permuted [7]) of each feature for all ML algorithms. Figure 4 shows that the importance of a feature varies with respect to the algorithm. For example, with **SVM** and **KNN**, some statistical features are more important than others, while with **RF** and **DT**, the features bearing the highest importance values are more equally distributed in each feature category. There are also features that bears negative importance values with algorithms, but not every time, while they always have positive importance values with other algorithms. There is no feature that always bears zero or negative importance values with one given algorithm, which means that all our features have a contribution to the ML classifiers. With **RF**, which bears the best performance, the most important feature is the location ratio. By checking the CSV files, we observe that most of the measures are situated at the last part of the file, while most of the columns in the front part are descriptive, which probably explains the importance of the location ratio.

---

[4] https://github.com/Implementation111/measure-detection.

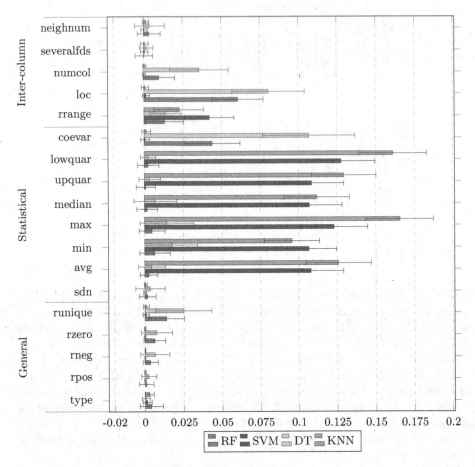

Fig. 4. Feature importance

5   Conclusion and Future Work

In this paper, we propose a machine learning-based method for detecting OLAP measures from tabular data. Our method is mainly dedicated to tabular data of simple structure, since the case of complex structures has been addressed in the literature. Some complex structures can also be converted to simple structures. To fuel machine learning algorithms, we define three categories of features for numerical columns in tabular data. We experiment with several real-world CSV datasets and test four machine learning algorithms, among which Random Forest performs the best. We also analyze how the features we build contribute to the results.

In the future, we aim at considering not only numerical measures, but also textual measures [23]. Moreover, we only use CSV data in our experiments. Complementing them with other types of tabular data, including complex-structured data that are found in data lakes [24], could be relevant.

Acknowledgements. The research depicted in this paper is funded by the French National Research Agency (ANR), project ANR-19-CE23-0005 BI4people (Business Intelligence for the people).

Appendix: Dataset Information

In this appendix, we detail the information about the datasets we use in our experiments. As mentioned in Sect. 4.1, these datasets come from different sources: **AFD**[5], **CDC**[6], **CA**[7], **FR**[8], **KG**[9], **NZ**[10], **UK**[11], **US**[12], **WB**[13]. files from **AFD** and **FR** are in French while the others are in English.

Table 3. Number of files by domains

Economy	Health	Government	Environment	Society
143	57	80	28	38

Table 4 shows information about each data source and all data sources (**Total**), including the number of files ($\mathbf{N}_f$), the number of numerical columns ($\mathbf{N}_c$), the number of measures ($\mathbf{N}_m$) and the ratio of number of measures by the number of numerical columns ($\mathbf{R}_m$). Figures in brackets are the minimums and maximums. The original datasets and even more information about them can be found in our github.

---

[5] https://opendata.afd.fr.
[6] https://data.cdc.gov.
[7] https://open.canada.ca.
[8] https://www.data.gouv.fr.
[9] https://www.kaggle.com.
[10] https://www.stats.govt.nz.
[11] https://data.gov.uk.
[12] https://www.data.gov.
[13] https://data.worldbank.org.

Table 4. Data source characteristics

	AFD[a]	CDC[b]	CA[c]	FR[d]	KG[e]
N_f	7	28	23	30	106
N_c (min-max)	15 (1–4)	100 (1–4)	156 (2–28)	123 (1–38)	394 (1–17)
N_m (min-max)	8 (0–3)	70 (1–6)	113 (0–28)	39 (0–7)	271 (0–10)
R_m(%)	53.33	70.00	72.44	31.71	68.78
	NZ[f]	UK[g]	US[h]	WB[i]	Total
N_f	22	42	71	17	346
N_c (min-max)	62 (1–13)	137 (1–9)	311 (1–20)	84 (1–18)	1382 (1–38)
N_m (min-max)	8 (0–3)	99 (0–8)	194 (0–18)	63 (0–13)	900 (0–28)
R_m(%)	69.35	72.26	62.38	75.00	65.12

[a] https://opendata.afd.fr.
[b] https://data.cdc.gov.
[c] https://open.canada.ca.
[d] https://www.data.gouv.fr.
[e] https://www.kaggle.com.
[f] https://www.stats.govt.nz.
[g] https://data.gov.uk.
[h] https://www.data.gov.
[i] https://data.worldbank.org.

Finally, the datasets that we choose contain at least one numerical column and can be used for DW creation. There were files that are used for other specific purpose, e.g., machine learning, which are not suitable to DW creation. They were thus discarded. There were also files with very poor data quality or completely lacking the information to understand the semantic meaning of columns, which made difficult to tell whether a column could be a measure. Such files were also discarded.

References

1. Abelló, A., et al.: Fusion cubes: towards self-service business intelligence. Int. J. Data Warehouse. Min. **9**(12), 66–88 (2013)
2. Adelfio, M.D., Samet, H.: Schema extraction for tabular data on the web. VLDB Endowment **6**(6), 421–432 (2013)
3. Alobaid, A., Kacprzak, E., Corcho, O.: Typology-based semantic labeling of numeric tabular data. Semant. Web **1**, 1–5 (2019)
4. Ballard, C., Herreman, D., Schau, D., Bell, R., Kim, E., Valencic, A.: Data modeling techniques for data warehousing
5. Chen, Z., Cafarella, M.: Automatic web spreadsheet data extraction. In: 3rd International Workshop on Semantic Search Over the Web, pp. 1–8 (2013)
6. Du, L., et al.: TabularNet: a neural network architecture for understanding semantic structures of tabular data. In: 27th ACM SIGKDD Conference on Knowledge Discovery & Data Mining, pp. 322–331 (2021)

7. Fisher, A., Rudin, C., Dominici, F.: All models are wrong, but many are useful: learning a variable's importance by studying an entire class of prediction models simultaneously. J. Mach. Learn. Res. **20**(177), 1–81 (2019)
8. Giorgini, P., Rizzi, S., Garzetti, M.: Goal-oriented requirement analysis for data warehouse design. In: 8th ACM International Workshop on Data Warehousing and OLAP, pp. 47–56 (2005)
9. Golfarelli, M., Rizzi, S., Vrdoljak, B.: Data warehouse design from xml sources. In: 4th ACM International Workshop on Data Warehousing and OLAP, pp. 40–47 (2001)
10. Golfarelli, M., Maio, D., Rizzi, S.: Conceptual design of data warehouses from E/R schemes. In: Proceedings of the Thirty-First Hawaii International Conference on System Sciences, vol. 7, pp. 334–343 (1998)
11. Golfarelli, M., Rizzi, S.: Data Warehouse Design: Modern Principles and Methodologies, 1st (edn.). McGraw-Hill Inc (2009)
12. Horner, J., Song, I.Y., Chen, P.P.: An analysis of additivity in OLAP systems. In: 7th ACM International Workshop on Data Warehousing and OLAP, pp. 83–91 (2004)
13. Song, I.-Y., Khare, R., Dai, B.: Samstar: a semi-automated lexical method for generating star schemas from an entity-relationship diagram. In: ACM 10th International Workshop on Data Warehousing and OLAP, pp. 9–16 (2007)
14. Jensen, M.R., Holmgren, T., Pedersen, T.B.: Discovering multidimensional structure in relational data. In: Kambayashi, Y., Mohania, M., Wöß, W. (eds.) DaWaK 2004. LNCS, vol. 3181, pp. 138–148. Springer, Heidelberg (2004). https://doi.org/10.1007/978-3-540-30076-2_14
15. Kimball, R.: A dimensional modeling manifesto. DBMS **10**(9), 58–70 (1997)
16. Koci, E., Thiele, M., Romero, O., Lehner, W.: A machine learning approach for layout inference in spreadsheets. In: International Joint Conference on Knowledge Discovery, Knowledge Engineering and Knowledge Management, pp. 77–88 (2016)
17. Lautert, L.R., Scheidt, M.M., Dorneles, C.F.: Web table taxonomy and formalization. ACM SIGMOD Rec. **42**(3), 28–33 (2013)
18. LeCun, Y., Bengio, Y., Hinton, G.: Deep learning. Nature **521**, 436–444 (2015)
19. Papenbrock, T., et al.: Functional dependency discovery: an experimental evaluation of seven algorithms. In: VLDB Endowment, vol. 8, pp. 1082–1093 (2015)
20. Papenbrock, T., Bergmann, T., Finke, M., Zwiener, J., Naumann, F.: Data profiling with metanome. VLDB Endowment **8**(12), 1860–1863 (2015)
21. Papenbrock, T., Naumann, F.: A hybrid approach to functional dependency discovery. In: International Conference on Management of Data, pp. 821–833 (2016)
22. Phipps, C., Davis., K.C.: Automating data warehouse conceptual schema design and evaluation. In: 4th International Workshop on Design and Management of Data Warehouses, pp. 23–32 (2002)
23. Ravat, F., Teste, O., Tournier, R., Zurfluh, G.: Top_keyword: an aggregation function for textual document OLAP. In: Data Warehousing and Knowledge Discovery, pp. 55–64 (2008)
24. Ravat, F., Zhao, Y.: Data lakes: trends and perspectives. In: Database and Expert Systems Applications, pp. 304–313 (2019)
25. Romero, O., Abelló, A.: A survey of multidimensional modeling methodologies. Int. J. Data Warehouse. Min. **5**(2), 1–23 (2009)
26. Romero, O., Abelló, A.: A framework for multidimensional design of data warehouses from ontologies. Data Knowl. Eng. **69**(11), 1138–1157 (2010)

27. Bimonte, S., Sautot, L., Journaux, L., Faivre, B.: Multidimensional model design using data mining: a rapid prototyping methodology. Int. J. Data Warehouse. Min. **13**(1), 35 (2017)
28. Sanprasit, N., Jampachaisri, K., Titijaroonroj, T., Kesorn, K.: Intelligent approach to automated star-schema construction using a knowledge base. Expert Syst. Appl. **182**, 115–226 (2021)
29. Sen, P.C., Hajra, M., Ghosh, M.: Supervised classification algorithms in machine learning: a survey and review. In: Emerging Technology in Modelling and Graphics, pp. 99–111 (2020)
30. Tryfona, N., Busborg, F., Borch Christiansen, J.G.: starER: a conceptual model for data warehouse design. In: Proceedings of the 2nd ACM International Workshop on Data Warehousing and OLAP, pp. 3–8 (1999)
31. Wang, Z., et al.: Tuta: tree-based transformers for generally structured table pre-training. In: 27th ACM SIGKDD Conference on Knowledge Discovery & Data Mining, pp. 1780–1790 (2021)
32. Yang, Y., Darmont, J., Ravat, F., Teste, O.: Automatic Integration Issues of Tabular Data for On-Line Analysis Processing. In: 16e journées EDA Business Intelligence & Big Data (EDA 2020), vol. B-16, pp. 5–18 (2020)

Discovering Overlapping Communities Based on Cohesive Subgraph Models over Graph Data

Said Jabbour[1]($\boxtimes$), Mourad Kmimech[2], and Badran Raddaoui[3]

[1] CRIL - CNRS UMR 8188, University of Artois, Arras, France
jabbour@cril.fr
[2] ISMM, University of Monastir, Monastir, Tunisia
[3] SAMOVAR, Télécom SudParis, Institut Polytechnique de Paris, Palaiseau, France
badran.raddaoui@telecom-sudparis.eu

Abstract. Detecting and analyzing dense subgroups or communities from social and information networks has attracted great attention over last decade due to its enormous applicability in various domains. A number of approaches have been made to solve this challenging problem using different quality functions and data structures. A number of cohesive structures have been defined as a primary element for community discovery in networks. Unfortunately, most of these structures suffer from computational intractability and they fail to mine meaningful communities from real-world graphs. The main objective of the paper is to exploit some cohesive structures in one unified framework to detect high-quality communities in networks. First, we revisit some existing subgraph models by showing their limits in terms of cohesiveness, which is an elementary aspect in graph theory. Next, to make these structures more effective models of communities, we focus on interesting configurations that are larger and more densely connected by fulfilling some new constraints. The new structures allow to ensure a larger density on the discovered clusters and overcome the weaknesses of the existing structures. The performance studies demonstrate that our approach significantly outperform state-of-the-art techniques for computing overlapping communities in real-world networks by several orders of magnitude.

Keywords: Graph mining · Overlapping community detection · k-truss · k-edge-connected component

1 Introduction

In various real-world domains, graphs are explored to represent data and their relationships such that nodes model the entities of interest and edges model the relationships between these different entities [6]. The distribution of edges in real graphs is often not uniform and it is often possible to discover highly connected nodes. The automatic detection of these groups of nodes, called *communities*, helps to identify some properties of real-world graph, such as sociology,

R. Wrembel et al. (Eds.): DaWaK 2022, LNCS 13428, pp. 189–201, 2022.
https://doi.org/10.1007/978-3-031-12670-3_16

bibliography, and biology, among others. Fundamentally, community detection aims to partition a graph into clusters, typically a group of vertices with more interactions amongst its members than the remainder of the network.

Various particular form of structures as communities in real-world networks have been developed in the past. The most basic structure is the triangle which is a clique of size 3. Since friends of friends tend to be friends themselves, many real-world networks contain a large number of triangles [10]. Obviously, one can consider maximal, w.r.t. set inclusion, cliques as communities in the graph. However, finding maximal cliques in large graphs is computationally intractable. Additionally, the clique structure is too restrictive. Hence, other more relaxed forms of cliques were proposed. Luce et al. introduced a distance-based model called k-clique [15], and Alba et al. proposed a diameter-based model called k-club [4]. Generally speaking, these models relax the reachability among vertices from 1 to k. However, they do not remove either the problems of enumeration or computational intractability. Other proposals focus on the degree constraint of the clique, like k-plex [11] and k-core [19]. Unfortunately, the k-plex enumeration problem is still NP-Complete since it restricts the subgraph size, while k-core is usually not powerful enough for uncovering the detailed community structure although it is computationally quite efficient [18]. Recently, a remedy to these limitations was the edge triangle based model, namely k-truss decomposition [20], is more suitable for social network analysis. In the same vain, the authors in [6] investigate the problem of discovering another structure that could approximately model communities in social networks, namely k-edge connected. Unfortunately, despite their structural properties, the aforementioned graph models are still looser than real communities in networks and the results are not that cohesive, i.e., the obtained groups might not be well-connected.

Herein, we focus in this paper on cliques, and the two widely used models of pseudo-cliques, that is, k-truss, and k-edge connected models. This paper's main contributions are briefly as follows. We propose to exploit these structures in one unified framework to detect overlapping communities in networks. To make these structures more effective models of communities, we focus on interesting configurations that are larger and more densely connected by fulfilling some triangle relationships. More precisely, we add additional vertices to the original structure, so the obtained subgraph is more densely connected. We also study some properties of the proposed structures. We stress here that the new structures are generic enough to encompass prior subgraph models. Lastly, extensive experiments on several real-world graphs show that the proposed algorithm outperforms existing techniques.

The remainder of the present paper is structured as follows. Section 2 introduces some preliminary definitions and notations. In Sect. 3, we present our novel cohesive subgraph models for overlapping communities discovery in large graphs. Experimental evaluation on several real-world graphs to show the performance of our proposed methods in Sect. 4. Finally, we draw conclusions in Sect. 5.

2   Formal Preliminaries

We provide in this section basic notions of overlapping community discovery problem and classical subgraph models used in the paper.

An undirected graph is a pair $G = (V, E)$ where V is a finite set of nodes ($n = |V|$) and $E \subseteq V \times V$ is the set of edges ($m = |E|$). A pair $G' = (V', E')$ is a *subgraph* of a given graph G if and only if $E' \subset E$ and $V' \subset V$. For a node $u \in V$, the set of adjacents of u is $\Gamma(u) = \{v \mid (u, v) \in E\}$. A triangle, denoted by Δ_{uvw} s.t. $u, v, w \in V$, is a cycle of length 3 in G. Given an edge $e = (u, v) \in E$, the support of e in G is the number of triangles that contain e, i.e., $Sup_G(e) = |\{w \in V \mid \Delta_{uvw} \in G\}|$.

Nodes in real graphs are structured into interconnected vertices, commonly coined *communities* or clusters. Note that these communities often overlap as nodes can be contained in different communities. The overlapping community discovery problem consists in splitting a network of interest into (overlapping) communities for intelligent analysis. It has recently attracted significant attention in diverse application domains. Finding network communities is, therefore, critical for characterizing the organizational structures and understanding complex systems. A great deal of effort has been devoted to developing overlapping community finding approaches (see [8,14] for an overview).

In networks, communities are generally associated with densely interconnected subgraphs. Indeed, a cohesive subgraph is an important vehicle for the detection of communities in real networks, such as biological networks, social networks and collaboration networks. The most intuitive definition for a cohesive subgraph is a clique in which each vertex is adjacent to every other vertex. More formally,

Definition 1 (Clique). *Let $G = (V, E)$ be a graph. Then, G is a clique if and only if $\forall u \in V$, $|\Gamma(u)| = |V| - 1$.*

Definition 2 (Maximal clique). *Let $G = (V, E)$ be a graph and $G' = (V', E')$ a clique in G. Then, G' is a maximal clique iff $\nexists G'' = (V'', E'')$ in G s.t. G'' is a clique and $V' \subset V''$. We denote by $M(G)$ the set of all maximal cliques in G.*

In real-world graphs, large communities hardly appear as cliques, and various dense subgraph models have been studied over the year. The most relevant ones that we adopt in this paper are k-truss, and k-edge connected models, which have gained popularity to model overlapping communities in real-world networks [2,3].

Definition 3 (k-truss). *Let $G = (V, E)$ be a graph and k be a positive integer s.t. $k \geq 3$. Then, G is a k-truss iff $\forall \, e \in E$, $Sup_G(e) \geq k - 2$.*

Intuitively, a k-truss is a maximal graph such that each pair of vertices within an edge has at least $k - 2$ common neighbors [7].

Definition 4 (k-edge connected). *Let $G = (V, E)$ be a graph. Then, G is a maximal k-edge connected iff $\forall u \in V$, $|\Gamma(u)| \geq k$ and G is connected when removed any $k - 1$ edges.*

Obviously, a maximal k-edge connected component is a connected neighborhood graph that cannot be disconnected by removing less than k edges.

Unfortunately, despite their structural properties, the previous explicit structures are not suitable to model real-world communities. To tackle this issue, we present in the next section a novel class of cohesive subgraph models for communities detection in networks.

3   Novel Cohesive Models for Community Discovery

In this section, we propose three novel dense subgraph models which leverage on a new type of cohesive components based on numerous existing structures. The underlying intuition is that all the aforementioned subgraph structures are a formal and strict way of defining accurate communities. Indeed, in the vast majority of real-world graphs, one can clearly distinguish groups of vertices that are highly connected. Furthermore, communities in social networks must be dense in terms of triangles. In fact, social networks are known to contain more triangles than expected by chance, which gives a community structure to the network. Such these important triangle relationships are somehow neglected in the aforementioned structures.

In light of the above, we present three novel cohesive models based on the principle that triangles are a good indicator of community structure. To do so, we first start with existing cohesive structures as "seeds" for growing final communities. For a good balance of our structures, we expand these seed sets using a seeding strategy. In words, our approach relies on two steps: the first step is to build initial communities that will be improved incrementally. These initial communities are modeled using either clique, k-truss or k-edge-connected structures. The second step is to grow clusters around seeds to refine the quality of final communities. This can be done by adding additional vertices to the initial communities while keeping the fundamental aspect of cohesiveness. Based on the notion of triangle, we have opted for the following principle: given an initial community C, any node of the original graph which forms at least n triangles with C must be added to C. In other words, if there are at least two vertices u and v in the initial community C such that u and v form a triangle with an external node w, then w must be added to the community, i.e., $C \cup \{w\}$. Of course, since vertices in social networks can often belong to multiple communities at once, the node w can be added to various subgroups.

Next, let us start with cliques as seeds. Communities based on large cliques are often undesirable as many of the returned clusters have large overlapping parts. This redundancy leads to various problems in both computational efficiency and usefulness of maximal cliques. So, as a first step, we aim at providing a concise and complete summary of the set of maximal cliques of a given network. For this, let us recall below the notion of visibility proposed in [21].

Definition 5 (Visibility). *Let $G = (V, E)$ be a graph and $S \subseteq M(G)$. The visibility of a maximal clique $C \in M(G)$ w.r.t. S, denoted by $V_S(C)$, is the maximum ratio of coverage of C by any $C' \in S$, i.e.,*

$$V_S(C) = max_{C' \in S} \frac{|C \cap C'|}{|C|}$$

Given a ratio $\tau \in [0, 1]$, the main idea is to compute the τ-maximal clique subset $S \subseteq M(G)$, such that the visibility of each maximal clique $C \in M(G)$ w.r.t. S is at least τ, i.e., $V_S(C) \geq \tau$ (see [21] for an overview). τ-MCS is short for τ-maximal clique subset from now on.

Based on the τ-MCS property, we are able to introduce our novel subgraph model that enhances cliques by paying attention to cohesiveness. In order to do so, we first define the following key concept:

Definition 6. *Let $G' = (V', E')$ be a subgraph of a graph $G = (V, E)$. Then, the support of a node $u \in V \setminus V'$ w.r.t. G', denoted as $Sup_G(u, G')$, is the number of triangles formed by u with G', that is, $Sup_G(u, G') = |\{\Delta_{uvw} \mid v, w \in V'\}|$.*

Based on Definition 6, we are now in a position to define the (τ, n)-MCS:

Definition 7. $((\tau, n)$**-MCS**). *Let $\tau \in [0, 1]$ and $n \geq 0$. A graph $G = (V, E)$ is (τ, n)-MCS if there exists a subgraph $G' = (V', E')$ of G such that:*

- *G' satisfies the τ-MCS property, and*
- *$\forall u \in V \setminus V'$, $Sup_G(u, G') \geq n$.*

Compared to maximal cliques, (τ, n)-MCS can significantly capture the interesting and important structural information out the scope of τ-MCS. We illustrate this in Fig. 1.

Example 1. Consider the undirected graph $G = (V, E)$ as shown in Fig. 1. Obviously, G has two maximal cliques $C_1 = \{1, 2, 3, 4, 5, 6, 7\}$ and $C_2 = \{5, 6, 7, 8, 9, 10\}$. Now, by setting $\tau = 0.3$, the τ-MCS of G is C_1 which is not suitable to directly model the whole graph as a cohesively connected community. By the definition of (τ, n)-MCS, the entire graph forms a single community for $n = 1$, since the vertices $\{8, 9, 10\}$ are contained in at least one triangle with C_1.

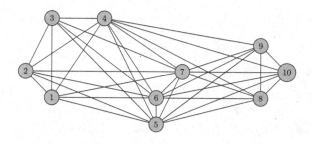

Fig. 1. A simple undirected graph

Now, despite the fact that a truss model is defined on the set of edges involved in various triangle relationships, it fails to identify accurate communities over graphs. Taking advantage of the triangle cohesiveness power, we propose a novel k-truss model, which we called (k,n)-truss, defined as following.

Definition 8 ((k,n)-truss). *Let n and k be two integers s.t. $n \geq 0$ and $k \geq 3$. Then, a graph $G = (V,E)$ is called a (k,n)-truss if there exists a subgraph $G' = (V',E')$ of G such that:*

- *G' is a k-truss, and*
- *$\forall\, u \in V \setminus V'$, $Sup_G(u,G') \geq n$.*

Intuitively, for the (k,n)-truss model, if two vertices have more common neighbors (i.e., $n \geq 1$), their relationship is stronger, which is overlooked in k-truss model. The following result shows the relation between the k-truss and (k,n)-truss models.

Proposition 1. *Let $n \geq 0$ and $k \geq 3$ be two integers. Then, the following results hold:*

(1) If $n = k$, then a (k,n)-truss is a $(k+2)$-truss.
(2) If $n = 0$, then a (k,n)-truss is a k-truss.
(3) If $n > 0$, then a (k,n)-truss is a $((k-1),(n-1))$-truss.

Clearly, Item 2 of Proposition 1 ensures that a k-truss is a particular case of the (k,n)-truss structure.

Example 2. Suppose the undirected graph represented by Fig. 2. This graph contains two 4-truss highlighted with gray and blue color. We can see that many vertices do not belong to these 4-truss despite their high connections. Now, if we consider the $(4,1)$-truss model, we can clearly see that the vertices highlighted with red color (resp. yellow color) are added to the blue (resp. gray) 4-truss, since they formed at least one triangle with the 4-truss.

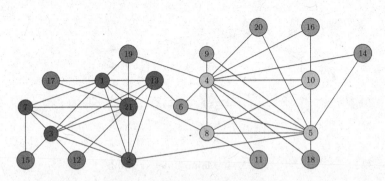

Fig. 2. An example of graph with 4-truss and $(4,1)$-truss models (Color figure online)

In the following, we revisit the k-edge connected structure in light of our triangle based cohesiveness strategy. So, on the basis of Definition 6, we introduce the (k, n)-edge connected structure as follows.

Definition 9 ((k, n)-edge connected). *A graph $G = (V, E)$ is called a (k, n)-edge connected with $k \geq 2$ and $n \geq 0$ if there exists a subgraph $G' = (V', E')$ of G such that:*

- *G' is a maximal k-edge connected, and*
- *$\forall u \in V \setminus V'$, $Sup_G(u, G') \geq n$.*

That is, the (k, n)-edge connected model is a generalization of k-edge connected with the proper parameter n. We would like to point out that n can make the graph size of (k, n)-edge connected flexible by changing its value. Thus, the following result holds.

Proposition 2. *Let $n \geq 0$ and $k \geq 2$ be two integers. Then, the following results hold:*

(1) If $n = 0$, then a (k, n)-edge connected is a k-edge connected.
(2) If $n \geq 1$, then a (k, n)-edge connected is a $((k - 1), n)$-edge connected.

The main advantage of a (k, n)-edge connected over a k-edge connected is its ability to gather closely related vertices in the same community.

Based on the novel cohesive models, we next design an algorithm to find overlapping communities in a graph by setting the parameters k, n and τ. The pseudo-code of the proposed algorithm is given by Algorithm 1. Given an input graph G, the set of all overlapping communities can be detected in two steps. The algorithm starts with the enumeration of seeds (Line 1). These seeds correspond to the initial communities. To do this, we apply an appropriate algorithm to enumerate the τ-MCS, the set of all k-truss or the maximal k-edge connected subgraphs. Once the detection of the set of seeds to the fixed values of parameters (i.e., k, n and τ) are performed, the algorithm tests for each seed G' the set of vertices outside G' that can be assigned to G' w.r.t. to the triangle based strategy. Specifically, each node that forms at least n triangles with the seed G' must be added to G'. This process is repeated iteratively until all the vertices outside the underlying seed are visited (Lines 3–9). The algorithm returns then all the final communities of G (Line 11).

Proposition 3. *Given an undirected graph G. Algorithm 1 returns exactly all the overlapping communities in G.*

4   Experimental Evaluation

As mentioned above, we propose to use our novel structures to detect overlapping communities in networks. We experimentally evaluate our approach, and show that the novel proposed subgraph models are efficient, and significantly improve upon the state-of-the-art algorithms.

Algorithm 1: Discovering Overlapping Communities via Subgraph Models

---

Data: $G = (V, E)$: an undirected graph, k, n: integers s.t. $k \geq 2$, $n \geq 1$,
     $\tau \in [0, 1]$: a ratio
Result: $\mathcal{C}$: the set of all communities in G
1 $\mathcal{S} \leftarrow$ Enumerate_Seeds(G);
2 $\mathcal{C} \leftarrow \mathcal{S}$;
3 **foreach** $G' = (V', E') \in \mathcal{S}$ **do**
4 | **foreach** $u \in V \setminus \mathcal{S}$ **do**
5 | | **if** $Sup_G(u, G') \geq n$ **then**
6 | | | $V' \leftarrow V' \cup \{u\}$
7 | | **end**
8 | **end**
9 | $\mathcal{C} \leftarrow \mathcal{C} \cup G'$
10 **end**
11 **return** $\mathcal{C}$

Our experimental evaluation was conducted on several networks that cover a variety of application areas and are briefly described in Table 1. All networks have ground-truth communities (see column 3 of Table 1). We have also evaluated three large scale graph data (i.e., `Amazon`, `DBLP`, and `Livejournal`) to demonstrate the scalability of our approach. We test the efficiency of our methods using the two well-known scoring metrics F1-score [23] and Normalized Mutual Information (NMI, for short) [12] defined by Eq. (1) and (2), respectively:

$$\frac{1}{2} \left(\frac{1}{|C^*|} \sum_{C_i \in C^*} F_1(C_i, \hat{C}_{g(i)}) + \frac{1}{|\hat{C}|} \sum_{\hat{C}_i \in \hat{C}} F_1(C_{g'(i)}, \hat{C}_i) \right) \qquad (1)$$

$$\frac{H(X) + H(Y) - H(X, Y)}{\frac{(H(X) + H(Y))}{2}} \qquad (2)$$

where $H(X)$ and $H(Y)$ are respectively the entropy of the variable X and Y in C' and C''. The joint entropy is denoted as $H(X, Y)$. When the two partitions C' and C'' are equal, this variable is set to 1.

We compare our method against the following most well-known state-of-the-art algorithms for discovering overlapping communities:

- *Clique Percolation Method* (CPM) [1]
- *Scalable Community Detection* (SCD) [17]
- *Communities from Edge Structure and Node Attributes* (CESNA) [24]
- *Cluster Affiliation Model for Big Networks* (BIGCLAM) [23]

For the CPM baseline, we consider the clique structure of size equal to 3.

Implementation Details: We implement (k, n)-truss, (k, n)-edge-connected subgraph and (τ, n)-MCS with the methods introduced in [20], [21] and [3], respectively. More precisely, we adapt and implement the methods in [20], [21] and [3] for our objective by taking into account the constraints required by the new structures. Our experiments were run on a PC with an Intel Core i5 processors with 8 GB memory. We imposed 4 h time-out for all the algorithms.

Table 1. Properties of the tested datasets

Network	Vertices/Edges	#Truth communities	Source
Karate	34/78	2	[22]
Dolphin	62/159	2	[16]
Railway	301/1224	21	[5]
Football	115/615	15	[9]
Book	105/441	3	[9]
Amazon	334863/925872	75149	[13]
DBLP	317080/1049866	13477	[13]
Live-journal	39979962/34681189	287512	[13]

4.1 Choosing the Best Value of Parameters

In our study, we consider several cohesive subgraph models. Note here that each of the existing structures has a parameter, i.e., k for k-truss and k-edge-connected and τ for τ-MCS. In fact, such parameters indicate how close the relationships are between vertices of the community. For the novel subgraph models, in addition to the previous parameters, we have to consider the additional parameter n to fix the number of external vertices to be added to the seeds. Our aim in this subsection is to select the best value of the parameters n, k and τ (i.e., giving the best value of the average F1-Score). To do that, we discover the communities on each considered network from Table 1, while varying k from 3 to 6 (for k-edge-connected, (k, n)-edge-connected, k-truss, and (k, n)-truss), and τ from 0.1 to 1 for the two cohesive subgraphs τ-MCS and (τ, n)-MCS. We also set n to 1 (the least restrictive condition under which the vertices are added). Figure 3 summarises the relationship between the average F1-score and the parameters for each cohesive subgraph model. We can clearly see from this figure that the best average of F1-score is obtained for $k = 4$ for the two cohesive subgraph models k-truss with a value of 0.420 and (k, n)-truss with a value of 0.481. Also, the best average of F1-score is obtained for $k = 3$ for the two cohesive subgraphs k-edge-connected with a value of 0.381, and (k, n)-edge-connected with a value of 0.395. For both the cohesive structures τ-MCS and (τ, n)-MCS, the best average F1-score is obtained for $\tau = 0, 3$. Interestingly, we can observe that the novel subgraph structures give the best values in terms of average F1-score compared to the existing models. This observation confirms our intuition that the new models of community discovery are more densely connected than the existing subgraphs.

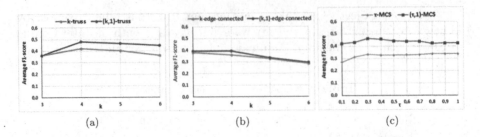

Fig. 3. Average F1-score for the different subgraph models with $n = 1$

Table 2. Comparison of the different cohesive subgraph models

Comparison based on F1-Score

Dataset	4-truss	(4, 1)-truss	3-edge-connected	(3, 1)-edge-connected	0.3-MCS	(0.3, 1)-MCS
Karate	0.523	**0.822**	0.572	**0.666**	0.374	**0.804**
Dolphin	0.404	**0.519**	0.642	0.642	0.281	**0.454**
Football	**0.735**	0.724	0.178	0.178	0.549	**0.689**
Railway	0.327	**0.333**	0.287	0.287	0.364	**0.432**
Book	0.644	**0.675**	0.560	0.560	0.197	**0.340**
Amazon	0.449	**0.458**	0.426	0.432	0.401	**0.441**
DBLP	**0.184**	0.186	0.176	0.177	**0.420**	0.416
Livejournal	0.094	**0.129**	0.209	**0.217**	0.096	**0.122**
Average	0.420	**0.481**	0.381	**0.395**	0.335	**0.462**

Comparison based on NMI

Dataset	4-truss	(4, 1)-truss	3-edge-connected	(3, 1)-edge-connected	0.3-MCS	(0.3, 1)-MCS
Karate	0.255	**0.364**	0.000	0.000	0.189	**0.488**
Dolphin	0.195	**0.258**	0.000	0.000	0.108	**0.225**
Football	**0.605**	0.540	0.000	0.000	0.375	**0.529**
Railway	0.193	**0.201**	0.105	0.105	0.099	**0.177**
Book	0.340	**0.399**	0.242	0.242	0.100	**0.145**
Amazon	0.250	**0.255**	0.369	**0.373**	0.133	**0.174**
DBLP	**0.146**	0.144	**0.154**	0.152	0.213	**0.212**
Livejournal	0.044	**0.068**	0.213	**0.224**	0.144	**0.162**
Average	0.253	**0.279**	0.135	**0.137**	0.170	**0.264**

4.2   Comparison with Existing Cohesive Models

Our aim here is to assess the accuracy of the detected communities based on
the proposed structures compared to the existing ones. To do that, we conduct
an experimental comparison of the different cohesive subgraphs studied in this
paper using the metrics F1-score and NMI. As shown in Fig. 3, the best value of
k is 4 for k-truss and (k, n)-truss, $k = 3$ for k-edge-connected and (k, n)-edge-
connected, and the best value of τ is 0.3 for τ-MCS and (τ, n)-MCS. Table 2 gives
the different results obtained on each considered network. From the results, we
can clearly see that the $(4, 1)$-truss model performs better than 4-truss in terms
of F1-score and NMI on 6 out of 8 networks. We can also observe that $(3, 1)$-
edge-connected and $(0.3, 1)$-MCS outperform, respectively, the 3-edge-connected
and 0.3-MCS models, in both F1-score and NMI. In addition to that, $(4, 1)$-
truss outperforms 4-truss by 15.52% in terms of average F1-score and 10.27%

in terms of average NMI. Besides, $(3,1)$-edge-connected outperforms 3-edge-connected by 3.67% in terms of average F1-score and 1.48% in terms of average NMI. Furthermore, the $(0.3,1)$-MCS model outperforms (0.3)-MCS by 37.91% and 55.29% in terms of average F1-score and NMI, respectively. Overall, the experiment results obtained in this subsection confirm that the $(4,1)$-truss model is able to find the best communities in real networks.

Afterwards, we have evaluated the performance of the $(4,n)$-truss model by varying the value of the parameter n from 1 to 5. Table 3 shows the relationship between n and the value of F1-score[1]. From Table 3, we can clearly see that $(4,1)$-truss has the highest score of F1-score among all the other $(4,n)$-truss $(1 < n \leq 5)$. The value of F1-score decreases as n grows. This can be explained by the fact that additional vertices around the seeds form at most one triangle with these seeds. These vertices are added to the final communities.

Table 3. Experiment results of $(4,n)$-truss by varying the value of n

Dataset	(4,1)-truss	(4,2)-truss	(4,3)-truss	(4,4)-truss	(4,5)-truss
Karate	**0,822**	0,523	0,523	0,523	0,523
Dolphin	**0,519**	0,427	0,404	0,404	0,404
Football	0,724	0,732	0,732	**0,735**	**0,735**
Railway	**0,333**	0,328	0,328	0,328	0,328
Book	**0,675**	0,647	0,649	0,648	0,644
Amazon	**0,458**	0,451	0,450	0,449	0,449
DBLP	**0,186**	0,112	0,087	0,043	0,021
Livejournal	**0,129**	0,089	0,050	0,29	0,012

4.3   Comparison with Baseline Algorithms

From the previous subsections, we can clearly observe that the $(4,1)$-truss model is the best one among all the cohesive subgraphs studied in this paper. Table 4 compares the $(4,1)$-truss based community detection with the most prominent state-of-the-art solutions for discovering overlapping communities. Regarding the F1-score, it is easy to see that $(4,1)$-truss outperforms the other four methods in 5 out of 8 networks. In details, $(4,1)$-truss outperforms every baseline with an interesting margin on Karate, Dolphin, Book, Amazon and Livejournal dataset, using F1-score. In terms of average F1-score, $(4,1)$-truss outperforms CESNA by 44.87%, SCD by 24.77%, CPM by 26.57%, and BIGCLAM by 89.37%. Regarding the NMI score, $(4,1)$-truss outperforms the baselines we have selected in 5 out of 8 networks. According to the average of NMI, $(4,1)$-truss outperforms CESNA by 34.13%, SCD by 38.71%, CPM by 49.19%, and BIGCLAM by 125%.

---

[1] Similar results have been seen for NMI metric.

Table 4. Comparison with baseline algorithms

Dataset	Comparison based on F1-score					Comparison based on NMI				
	(4, 1)-truss	CESNA	SCD	CPM	BIGCLAM	(4, 1)-truss	CESNA	SCD	CPM	BIGCLAM
Karate	**0.822**	0.046	0.574	0.439	0.369	0.364	0.231	**0.366**	0.216	0.182
Dolphin	**0.519**	0.284	0.311	0.404	0.206	**0.258**	0.137	0.145	0.195	0.081
Football	0.724	**0.754**	0.695	0.365	0.611	0.540	**0.777**	0.420	0.223	0.436
Railway	0.333	**0.414**	0.340	0.326	0.390	**0.201**	0.165	,084	0.120	0.138
Book	**0.675**	0.512	0.452	0.538	0.201	**0.399**	0.332	0.282	0.280	0.105
Amazon	**0.458**	0.161	0.388	0.454	0.153	**0.255**	0.003	0.158	0.219	0.032
DBLP	0.186	0.059	0.303	0.413	0.091	0.144	0.003	0.146	0.220	0.031
Livejournal	**0.129**	0.018	0.096	0.102	0.013	**0.068**	0.008	0.016	0.022	0.002
Average	**0.481**	0.332	0.395	0.380	0.254	**0.279**	0.208	0.202	0.187	0.124

5   Conclusion

In this work, we focused on detecting overlapping communities in real-world networks. We first provided a detailed review of previous work in cohesive subgraph models. Then, we proposed three new effective structures to model communities in networks. The idea is to focus on interesting configurations that are larger and more densely connected than existing structures. We further improved the performance of the new structures by detecting the overlapping communities in networks.

There are two open questions that might be worth to look at. The first is about the parameters of the new structures. Possible improvements can be obtained by designing better parameters to select additional vertices. Second, we believe that an extension of our proposed approach to handle dynamic community detection in networks can be more helpful in real-life applications.

References

1. Adamcsek, B., Palla, G., Farkas, I.J., Derényi, I., Vicsek, T.: Cfinder: locating cliques and overlapping modules in biological networks. Bioinformatics **22**(8), 1021–1023 (2006)
2. Akbas, E., Zhao, P.: Truss-based community search: a truss-equivalence based indexing approach. Proc. VLDB Endow. **10**(11), 1298–1309 (2017)
3. Akiba, T., Iwata, Y., Yoshida, Y.: Linear-time enumeration of maximal k-edge-connected subgraphs in large networks by random contraction. In: CIKM, pp. 909–918 (2013)
4. Alba, R.D.: A graph-theoretic definition of a sociometric clique. J. Math. Sociol. **3**, 113–126 (1973)
5. Chakraborty, T., Srinivasan, S., Ganguly, N., Mukherjee, A., Bhowmick, S.: On the permanence of vertices in network communities. In: SIGKDD, pp. 1396–1405 (2014)
6. Chang, L., Yu, J.X., Qin, L., Lin, X., Liu, C., Liang, W.: Efficiently computing k-edge connected components via graph decomposition. In: SIGMOD, pp. 205–216 (2013)

7. Fang, Y., Huang, X., Qin, L., Zhang, Y., Zhang, W., Cheng, R., Lin, X.: A survey of community search over big graphs. VLDB J. **29**, 1–40 (2019). https://doi.org/10.1007/s00778-019-00556-x

8. Fortunato, S.: Community detection in graphs. CoRR abs/0906.0612 (2009)

9. Girvan, M., Newman, M.E.J.: Community structure in social and biological networks. Proc. Natl. Acad. Sci. **99**(12), 7821–7826 (2002)

10. Jabbour, S., Mhadhbi, N., Raddaoui, B., Sais, L.: Triangle-driven community detection in large graphs using satisfiability. In: AINA, pp. 437–444 (2018)

11. Jabbour, S., Mhadhbi, N., Raddaoui, B., Sais, L.: A declarative framework for maximal k-plex enumeration problems. In: AAMAS (2022)

12. Lancichinetti, A., Fortunato, S., Kertesz, J.: Community detection algorithms: a comparative analysis. New J. Phys. **11** (2009)

13. Leskovec, J., Krevl, A.: SNAP Datasets: stanford large network dataset collection. http://snap.stanford.edu/data (2014)

14. Leskovec, J., Lang, K.J., Mahoney, M.W.: Empirical comparison of algorithms for network community detection. In: WWW, pp. 631–640 (2010)

15. Luce, R.D.: Connectivity and generalized cliques in sociometric group structure. Psychometrika **15**(2), 169–190 (1950). https://doi.org/10.1007/BF02289199

16. Lusseau, D., Schneider, K., Boisseau, O., Haase, P., Slooten, E., Dawson, S.: The bottlenose dolphin community of doubtful sound features. Behav. Ecol. Sociobiol. **54**(4), 396–405 (2003). https://doi.org/10.1007/s00265-003-0651-y

17. Prat-Pérez, A., Dominguez-Sal, D., Larriba-Pey, J.: High quality, scalable and parallel community detection for large real graphs. In: WWW, pp. 225–236 (2014)

18. Saito, K., Yamada, T., Kazama, K.: Extracting communities from complex networks by the k-dense method. IEICE Transactions **91**-**A**(11), 3304–3311 (2008)

19. Seidman, S.B.: Network structure and minimum degree. Soc. Netw. **5**(3), 269–287 (1983)

20. Wang, J., Cheng, J.: Truss decomposition in massive networks. In: ACM (2013)

21. Wang, J., Cheng, J., Fu, A.W.C.: Redundancy-aware maximal cliques. In: ACM (2013)

22. Zachary, W.W.: An information flow model for conflict and fission in small groups. J. Anthropol. Res. **33**, 452–473 (1977)

23. Yang, J., Leskovec, J.: Overlapping community detection at scale: a nonnegative matrix factorization approach. In: WSDM, pp. 587–596 (2013)

24. Yang, J., McAuley, J.J., Leskovec, J.: Community detection in networks with node attributes. In: ICDM, pp. 1151–1156 (2013)

Discovery of Keys for Graphs

Morteza Alipourlangouri$^{(\boxtimes)}$ and Fei Chiang

McMaster University, 1280 Main St W, Hamilton, ON L8S 4L8, Canada
{alipoum,fchiang}@mcmaster.ca

Abstract. Keys for graphs specify the topology and value constraints to uniquely identify entities in a graph in applications such as object identification, knowledge fusion, deduplication, and social network reconciliation. Despite their prevalence, existing key mining algorithms do not consider graph keys with recursive key definitions, which capture dependence between entities. We introduce GKMiner, an algorithm that mines recursive keys over graphs. We show the efficiency and utility of our discovered keys using large-scale, real data graphs.

Keywords: Graphs · Key · Knowledge graphs

1 Introduction

Keys uniquely identify an entity in applications such as deduplication, and entity linking [5]. While keys have been extensively used in relational settings, recent work has proposed keys for graphs such as PG-Keys [3], keys based on uniqueness constraints [9], and GKeys [4], which include sub-entities, and their associated (recursive) key definitions.

Despite their prevalence and utility, manual specification of keys is infeasible over increasingly large graphs. Existing key mining algorithms for graphs include RDF triples, and variants for approximate and conditional keys such as SAKey [10], and VICKEY [11], respectively, that prune the large space of candidates by returning only keys satisfying desirable support, and conditional constants. However, the discovered keys do not include sub-entities within its key definition (e.g., the key for a university relies on a key for its city), and do not permit specification of topological constraints. Consider the following example from DBpedia that highlights these facets.

Example 1. Figure 1 shows a sample of the DBpedia knowledge graph [6] of three colleges, five cities, and three countries. Consider graph keys with patterns P_1 and P_2 to identify college shown in Fig. 1. P_1 uses *name* and *motto*, and P_2 uses *name* and *city*. Note that city is a sub-entity, with its own unique key, that is recursively defined with respect to *city*), which uses name and country (P_3). Similarly, *country* is a recursive key defined using *name* (P_4). P_2 (resp. P_3) is dependant to P_3 (resp. P_4), which reflects the recursiveness of graph keys [4].

© The Author(s), under exclusive license to Springer Nature Switzerland AG 2022
R. Wrembel et al. (Eds.): DaWaK 2022, LNCS 13428, pp. 202–208, 2022.
https://doi.org/10.1007/978-3-031-12670-3_17

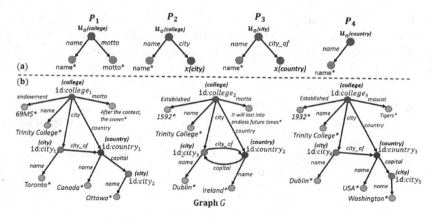

Fig. 1. Sample graph from DBpedia.

The above example highlights two points: (1) several keys are possible for an entity, with varying topology and attributes. This requires defining key properties to identify *frequent* (well-supported) keys that are *simple* (involving the fewest number of literals). Similar notions have been explored in the discovery of relational keys. (2) The scale and complexity of existing property graphs exhibit embedded entity relationships, and requirements to capture not only attribute values, but structural requirements as part of the key definition. This necessitates new techniques for mining graph keys to include the topology and the ability to consider recursive keys. In this paper, we propose GKMiner, a new algorithm that mines frequent, minimal graph keys, including recursive keys. GKMiner builds upon our earlier work to include a new support metric, and optimizations to improve the efficiency of the discovery process.

Contributions. (1) We define new properties for graph keys (*support* and *minimality*), and formalize the graph key discovery problem. (2) We introduce GKMiner, an algorithm that mines all recursive graph keys by using novel auxiliary structures and optimizations to prune unlikely key candidates. (3) Lastly, we evaluate GKMiner over real graphs, and show its scalability and efficiency.

2   Preliminaries

Graphs and Graph Patterns. A *directed graph* $G = (V, E, L, F)$ where: (i) V is a finite set of vertices; (ii) $E \in V \times L \times V$, *i.e.*, $e = (u, l, v)$ represents an edge from u to v with the label $l \in L$; (iii) L a finite set of labels; (iv) For a node v, $F(v)$ is a tuple specifying the set of attributes as $(A_1 = a_1, ..., A_n = a_n)$ of v. Each $v \in V$ may have a label $l \in L$ referred as v.type and a numeric id, denoted by v.id. A *graph pattern* is a connected, directed graph $P(u_o) = (V_P, E_P, L_P)$ where (1) V_P is a finite set of pattern nodes; (2) E_P is a finite set of pattern edges; (3) L_P is a function which assigns a specific label $L_P(v)$ (resp. $L_P(e)$) to each vertex $v \in V_P$ (resp. each edge $e \in E_P$). Pattern nodes V_P may be: (1) a *center* node $u_o \in V_P$, representing the key entity to mine; (2) a set of *variable nodes* $V_x \subseteq V_P$ that map to an entity along with a type and an id; and (3) a set of *constant nodes* $V_c = V_P \setminus (\{u_o\} \cup V_x)$ containing only constant values.

Graph Pattern Matching. Given two labels ι and ι' from L_P, we say ι matches ι', denoted as $\iota \asymp \iota'$ if either $\iota = \iota'$ or $\iota = `\_'$, *i.e.*, wildcard matches any label. Given a graph G and a pattern $P(u_o)$, a match h is a subgraph $G' = (V', E', L', F'_A)$, which is isomorphic to P, *i.e.*, there exists a bijective function h from V_P to V' such that (i) for $v \in V_P$, $L_P(v) \asymp L'(h(v))$; and (ii) for each edge $e(u, u') \in E_P$, there exists an edge $e'(h(u), h(u')) \in G'$ such that $L_P(e) \asymp L'(e')$.

Graph Keys (GKeys). A graph key consists of a pattern $P(u_o)$ for entity u_o [4]. Given two matches h_1 and h_2 of $P(u_0)$ in G, (h_1, h_2) satisfies $P(u_0)$ denoted as $(h_1, h_2) \models P(u_0)$, if (a)$\{\forall v \in V_x, h_1(v).\text{id} = h_2(v).\text{id}\}$; (b) $\{\forall v \in V_c, L(h_1(v)) \asymp L(h_2(v))\}$; and (c)$\{\forall e \in E_P, L(h_1(e)) = L(h_2(e))\}$; then $h_1(u_o).\text{id} = h_2(u_o).\text{id}$. A graph G satisfies key $P(u_o)$, denoted as $G \models P(u_o)$, if for every pair of matches $(h_1, h_2) \models P(u_0)$. Moreover, key $P(u_o)$ is a *recursive* key if it contains at least one variable node $v \neq u_o$, otherwise, $P(u_o)$ is called a *value-based* key [4]. In Fig. 1, $P_1(college)$ uniquely identifies $college_1$ and $college_2$ with different *motto*, despite having the same *name*. $P_2(college)$ is a recursive GKey, dependent on *city* of $P_3(city)$, and *country* of $P_4(country)$.

3   Discovering Graph Keys

We introduce two desirable properties of graph keys, formalize the graph key mining problem, and then present the GKMiner algorithm and its optimization.

3.1   Key Properties

GKey Embedding. We say a GKey $P(u_o) = (V_P, E_P, L_P)$ is *embeddable* in another GKey $P'(u_o) = (V'_P, E'_P, L'_P)$, if there exists a subgraph isomorphic mapping f from V_P to a subset of nodes in V'_P that preserves node labels/values of V_P, and all the edges that are induced by V_P with the corresponding edge labels.

Minimality. A GKey $P(u_o)$ is minimal if there exists no GKey $P'(u_o)$ such that $P'(u_o)$ is embeddable in $P(u_o)$. A set Σ of GKeys with $G \models \Sigma$ is minimal, if it does not contain any redundant GKeys. A redundant GKey $P(u_o)$ exists in Σ, if removing $P(u_o)$ from Σ results in a Σ' that is logically equivalent to Σ, *i.e.*, Σ' uniquely identifies the same entities as Σ in G.

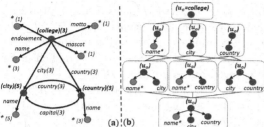

Table 1. Comparative accuracy.

Type	GKMiner	SAKey
	P/R/F1	P/R/F1
Book	0.99/**0.07/0.13**	1/0.03/0.06
Actor	1/**0.36/0.52**	0.99/0.27/0.43
Museum	1/**0.21/0.34**	1/0.12/0.21
Movie	**0.99/0.12/0.21**	0.99/0.04/0.08

Fig. 2. (a) $\mathcal{S}$; (b) $\mathcal{L}$ with sup = 75%

Support. We define support of a GKey $P(u_o)$ as the number of unique entities identified by $P(u_o)$ such that $G \models P(u_o)$ (denoted as $|P(u_o)|$) over the total number of instances of type u_o in G (denoted as N), i.e.,, $\mathsf{sup}(P(u_o)) = \frac{|P(u_o)|}{N}$.

k-bounded GKeys. For a given user defined natural number k, a GKey $P(u_o)$ is _k-bounded_ if $\mathsf{size}(P(u_o)) \leq k$, where $\mathsf{size}(P(u_o)) = |E_p| + \mathsf{size}(P(v_P))\forall v_p \in V_P$. A set Σ of GKeys is _k-bounded_, if each $P(u_o) \in \Sigma$ is _k-bounded_.

Problem Statement. Given a graph G, a type u_o, a support threshold δ, and a natural number k, mine all minimal _k-bounded_ GKeys Σ of the node type u_o, such that for each GKey $P(u_o) \in \Sigma$, $P(u_o)$ has the minimum support δ in G.

3.2 GKMiner Algorithm

GKMiner solves the aforementioned problem, we introduce its components next.

Summary Graph. We create an auxiliary structure, called a _summary graph_ $\mathcal{S}$ where for each node type in G, there exists a node in $\mathcal{S}$. An edge exists between two vertices $u_1, u_2 \in \mathcal{S}$ if there exists at least one edge between two entities of the same type(s) in G. For each node $u_1 \in \mathcal{S}$, we define an attribute u_1.count that computes the total number of entities in G of type u_1. Similarly, for each edge $e = (u_1, u_2)$, we define an edge counter, e.count, which sums all edges in G with end points of type u_1 and u_2.. We use $\mathcal{S}$ to prune all edges $e = (u_1, u_2)$ where u_2.count $< (\delta \cdot e$.count $)$. From the vertices in $\mathcal{S}$, we further prune attributes with frequency counts less than δ. The resulting $\mathcal{S}$ thus contains nodes, edges, and attributes $\mathcal{A} = \{A_1, \ldots, A_n\}$ with frequency counts of at least δ. For example, Fig. 2(a) shows the summary graph for Fig. 1, and the frequency counts for each node type. Of the three colleges, one has the attribute _endowment_, and all three have attribute _name_. Lastly, we create a candidate set $\mathcal{V} = \{v_1, \ldots, v_n\}$ of variable nodes in $\mathcal{S}$ connected to u_o to facilitate candidate key generation next. We refer the reader to our extended paper for details [2].

Lattice $\mathcal{L}$. To generate key candidates, we construct and traverse a lattice $\mathcal{L}$. At level $l = 0$, the root of $\mathcal{L}$ is a node x of type u_o. For each level $l, 1 \le l \le k$, we create a node in $\mathcal{L}$ representing all l-combinations of attributes and variable nodes drawn from $\mathcal{A}$ and $\mathcal{V}$, respectively. We add an edge from x to each node at level l, representing a candidate GKey. Candidate GKey $P(u_o)$ at level $l - 1$ is connected via an edge to $P'(u_o)$ at level l, if $P(u_o)$ is embedded in $P'(u_o)$. Figure 2(b) shows an example $\mathcal{L}$, where the last leaf node represents a candidate GKey with one attribute and two variables nodes.

GKMiner Algorithm and Optimizations. We introduce GKMiner, a sequential GKey mining algorithm that traverses lattice $\mathcal{L}$ level-wise to generate and evaluate candidate GKeys (represented as nodes in $\mathcal{L}$). We seek k-bounded GKeys that satisfy the given support level δ. We implement two pruning strategies for greater efficiency. First, we discover k-bounded GKeys, which is guaranteed by construction of $\mathcal{L}$ for candidate keys without variable nodes. A variable node may introduce a recursive key, and we must then check that $\mathsf{size}(P(u_o)) \le k$, and if not, prune such candidates. Second, for all discovered keys $(P(u_o))$, we prune all its descendant nodes in $\mathcal{L}$ for minimality. We further optimize GKMiner by identifying attributes A_i from nodes $v \in \mathcal{S}$ with (high) cardinality that would serve as (pseudo) key attributes. We maintain a hashmap for each attribute domain of A_i, and only validate entity matches of $P(u_o)$ where duplicate values in A_i occur. Clearly, if the cardinality of A_i is equal to the number of matches, we have discovered a key attribute. We present the algorithm pseudo-code in [2].

Handling Recursive GKeys. A novel aspect of GKMiner is the discovery of *recursive* keys. For a variable node of type t_1, GKMiner is recursively invoked to generate a lattice $\mathcal{L}(t_1)$ for candidate key generation. The challenge of handling recursive dependencies arises when keys for type u_o are dependent on types $t_1, t_2, \ldots t_z$ which is then dependent on u_o, thereby creating a cyclic dependency. We handle such cases by managing an auxiliary *dependency graph*, $\mathcal{D}$ (V_D, E_D) that creates a vertex for each dependent type t_1, and an edge exists between two vertices $(t_1, t_2) \in E_D$ if there exists a key dependence between their types. We implement a cycle prevention strategy, similar to deadlocks prevention by removing the last edge that created the cycle from $\mathcal{L}(t_i)$ [8].

4   Experiments

Setup. We implement our algorithms[1] in Java v17, using a Linux machine with AMD 2.7 GHz CPU and 128 GB of RAM. We used two real graphs: (1) DBpedia [6] with 5.04 M entities, 13.3 M edges, 421 and 584 distinct entity types and labels, respectively; (2) IMDB [1] with 6.1 M entities, 21.3 M edges, with 7 types (for brevity, results for IMDB are similar to DBpedia and can be found in [2]); (3)DBpediaYago [11] with ground truth of linking entities between DBpedia [6]

---

[1] https://github.com/mac-dsl/GraphKeyMiner.git.

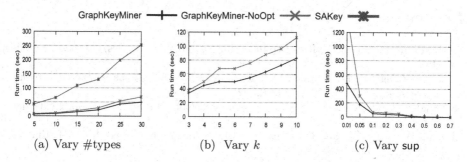

Fig. 3. GKMiner efficiency and effectiveness.

and Yago [7] with 6 types. Unless otherwise noted, we set default values of sup = 10%, $k = 5$, and 30 and 7 node types over DBpedia and IMDB, respectively.

Baselines. We evaluate GKMiner with optimizations (Sect. 3) against two baselines: (1) GKMiner-NoOpt, GKMiner without optimizations; and (2) SAKey [10].

Exp-1: Vary #types. Figure 3a shows the runtime for varying number of types on DBpedia. GKMiner is on average 30% faster than GKMiner-NoOpt, and 6 times faster than SAKey, demonstrating the effectiveness of our optimizations.

Exp-2: Vary k. Figure 3b shows that runtimes increase for increasing k over DBpedia due to the increased complexity of graph matching. On average, GKMiner runs 33% faster than GKMiner-NoOpt.

Exp-3: Vary sup. Figure 3c shows, as expected, runtimes decrease for increasing support due to more aggressive pruning of key candidates. On average, GKMiner runs 66% faster than GKMiner-NoOpt.

Exp-4: Effectiveness of GKMiner. We evaluate the quality of the keys from GKMiner against those from SAKey in an entity linking task using DBpediaYago with a given ground truth [11]. Table 1 shows that both techniques produce keys with high precision (over 98%), but low recall (less than 40%) due to data incompleteness. However, our discovered recursive keys lead to an average 7% and 9% gain in recall and F_1-score, respectively, over SAKey. The results over more types are reported in [2].

5 Conclusion and Future Work

We introduced GKMiner, an algorithm that discovers recursive keys over graphs. Our evaluations highlight the scalability of our techniques, the benefits of our optimizations, and the quality of our discovered keys. We intend to study the discovery of conditional GKeys, and a parallel discovery algorithm for GKeys.

References

1. Imdb dataset (2021). ftp://ftp.fu-berlin.de/pub/misc/movies/database/frozend ata/
2. Alipourlangouri, M., Chiang, F.: Discovery of keys for graphs [extended version]. arXiv preprint arXiv:2205.15547 (2022)
3. Angles, R., et al.: Pg-keys: keys for property graphs. In: SIGMOD, pp. 2423–2436 (2021)
4. Fan, W., Fan, Z., Tian, C., Dong, X.L.: Keys for graphs. Proceed. VLDB Endowment 8(12), 1590–1601 (2015)
5. Ilyas, I.F., Chu, X.: Trends in cleaning relational data: consistency and deduplication. Found. Trends Databases 5(4), 281–393 (2015)
6. Lehmann, J., Isele, R., Jakob, M., et al.: Dbpedia-a large-scale, multilingual knowledge base extracted from wikipedia. Semant. web 6(2), 167–195 (2015)
7. Mahdisoltani, F., Biega, J., Suchanek, F.: Yago3: a knowledge base from multilingual wikipedias. In: CIDR (2014)
8. Peterson, J.L., Silberschatz, A.: Operating System Concepts. Addison-Wesley Longman Publishing Co., Inc. (1985)
9. Skavantzos, P., Zhao, K., Link, S.: Uniqueness constraints on property graphs. In: La Rosa, M., Sadiq, S., Teniente, E. (eds.) CAiSE 2021. LNCS, vol. 12751, pp. 280–295. Springer, Cham (2021). https://doi.org/10.1007/978-3-030-79382-1_17
10. Symeonidou, D., Armant, V., Pernelle, N., Saïs, F.: SAKey: scalable almost key discovery in RDF data. In: Mika, P. (ed.) ISWC 2014. LNCS, vol. 8796, pp. 33–49. Springer, Cham (2014). https://doi.org/10.1007/978-3-319-11964-9_3
11. Symeonidou, D., Galárraga, L., Pernelle, N., Saïs, F., Suchanek, F.: VICKEY: mining conditional keys on knowledge bases. In: d'Amato, C. (ed.) ISWC 2017. LNCS, vol. 10587, pp. 661–677. Springer, Cham (2017). https://doi.org/10.1007/978-3-319-68288-4_39

OPTIMA: Framework Selecting Optimal Virtual Model to Query Large Heterogeneous Data

Chahrazed B. Bachir Belmehdi[1]([✉]), Abderrahmane Khiat[2,3],
and Nabil Keskes[1]

[1] ESI-SBA Institute, Sidi Bel Abbès, Algeria
{cb.bachirbelmehdi,n.keskes}@esi-sba.dz
[2] Fraunhofer IAIS, Enterprise Information Systems, Sankt Augustin, Germany
abderrahmane.khiat@iais.fraunhofer.de
[3] Fraunhofer Research Center for Machine Learning, Munich, Germany

Abstract. OPTIMA is a framework that enables querying the original data on-the-fly without any materialization. It implements two different virtual data models, GRAPH and TABULAR, to join and aggregate data. OPTIMA leverages ontology-based data access and calls the deep learning method to predict the optimal virtual data model using the features extracted from SPARQL queries. Extensive experiments show a reduction in query execution time of over 40% for the TABULAR model selection, and over 30% for the GRAPH model selection. **OPTIMA is available on GitHub** https://github.com/chahrazedbb/OPTIMA.

Keywords: Data virtualization · OBDA · Big data · Deep learning

1 Introduction

Data virtualization approaches tackle data integration challenges by creating a virtual data model under which the heterogeneous formats are homogenized *on-the-fly* without data materialization [4]. Ontology-Based Data Access (OBDA) [5] approaches maintain data virtualization with practical knowledge representation models and ontology-based mappings. Existing solutions [1–3] use by design a fixed model e.g., TABULAR as the only virtual data model[1] to load and transform the requested data into a uniform model to be joined.

Nevertheless, TABULAR virtual model can have downsides performances for queries that involve many join operations on large data. While other data models such as GRAPH perform better for such queries. On the other hands, the TABULAR model performs better for queries that involve selection or projection. Therefore, there is a need to support different virtual models and select

---

[1] We denote GRAPH and TABULAR to distinguish between Virtual model and data source models.

© The Author(s), under exclusive license to Springer Nature Switzerland AG 2022
R. Wrembel et al. (Eds.): DaWaK 2022, LNCS 13428, pp. 209–215, 2022.
https://doi.org/10.1007/978-3-031-12670-3_18

the optimal one depending on query behavior, thus saving operational execution time. In many cases it hard to predict the optimal virtual model since the selection depends on many criteria such query plan, data model, size, operations.

To address this challenge, we develop OPTIMA, an extensible framework that uses two Virtual model GRAPH and TABULAR and supports out-of-the-box five data models sources property Graph, Relational, Tabular Document-based and Wide-Columnar. To select the optimal virtual data model GRAPH or TABULAR, we used one hot vector encoding to transform different SPARQL features into hidden representations. Next, we embed these representations into a tree-structured model, which we use to classify the virtual model GRAPH or TABULAR that has the lowest query execution time. Experiments show that OPTIMA reduces query execution time of over 40% for the TABULAR model selection, and over 30% for the GRAPH model selection. We describe each component of our system OPTIMA as illustrated in Fig. 1; followed by conducted experiment and related work.

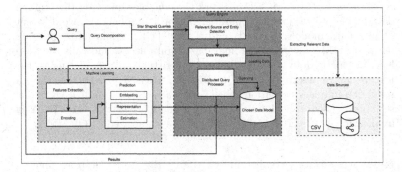

Fig. 1. OPTIMA architecture

2   OPTIMA: Optimal Virtual Model for Querying Large Heterogeneous Data

2.1   Virtual Data Model Prediction

Built on top of OBDA components, this distinctive component implemented in OPTIMA aims to select the optimal virtual data model GRAPH or TABULAR based on the query behavior. The component receives as input the SPARQL query and predicts the optimal virtual data model that has that the lowest execution time. The deep learning model starts first by breaking down the SPARQL query plan into nodes. Each node includes a set of query features that significantly affect the query execution time (e.g., filter). The different features are then encoded using one-hot vector. Next, we propose tree structured model that takes as input the encoded features of SPARQL query to learn the representation of each sub-plan effectively and outputs the optimal virtual data model,

GRAPH or TABULAR that has the lowest query execution time. Our model consists of an embedding layer to condense the features' vectors and an estimation layer to estimate the optimal virtual data model. In addition, the model includes an intermediate representation layer to capture the correlation between the joined star-shaped queries. Once the optimal model is predicted, the rest of the OBDA components and operations (e.g., join) follow the optimal virtual model predicted GRAPH or TABULAR.

2.2 Query Decomposition and Relevant Entity Detection

This component decomposes the SPARQL query it into star-shaped queries. More precisely, the query's Basic Graph Pattern (BGP) is divided into set of sub-BGPs, where each sub-BGP contains all the triple patterns sharing the same subject variable. Those sub-BGPs sharing the same subject are called star-shaped query. Next, this component analyzes each star-shaped query and visits mappings file to obtain the data source's path as well as the attributes that are mapped to each element of the star-shaped query i.e., relevant entities. This information then is passed to data wrapper to load relevant entities.

2.3 Data Wrapper

Once the sources and relevant entities are identified using mappings, data wrapper converts relevant entities (e.g., tables) from their original models to data that comply with optimal virtual data model predicted which is actually the data structure of the computation unit of the query engine. This conversion occurs at query-time, which allows for the parallel execution of expensive operations, e.g., join. Query engines implement already wrappers called connectors to convert data entities from the source to virtual data model, performing transformation of data source e.g., relational model to virtual model e.g., GRAPH (see Fig. 2a).

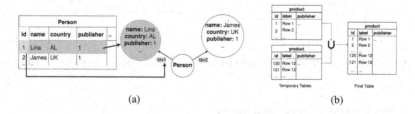

(a)   (b)

Fig. 2. (a) Transformation relational to GRAPH (b) Union temporary virtual TABULARs

Each star-shaped query corresponds to one relevant entity and thus one single virtual data model is created. This is the case when the relevant entity according to the mapping could be retrieved only from one data source. Otherwise, if the relevant entity according to the mapping could be retrieved from multiple data sources, then the virtual model for one relevant entity is the union of temporary virtual model created for each source (see Fig. 2b). Below we describe the data sources' model transformation by wrappers into GRAPH and TABULAR.

- For the virtual data model of type GRAPH, the result of the star-shaped query on Tabular and Relational models defined by CSV and MySQL respectively is a table with specific columns. A virtual GRAPH is created form the table. For each row of the table, a vertex is created that has the same label as the table's name (e.g., table 'Person' corresponds to all vertices with label "Person") in addition to the root vertex. Edges are created between vertices and the root vertex whereas the properties of each vertex are the columns of the table (e.g., column 'name' corresponds to property 'name') and the values of the properties are the table's cell information. Same process is applied for property graph defined by neo4j, document-based and Wide-Column models (e.g., JSON or XML file) defined by Cassandra and MongoDB.
- As for virtual data model of type TABULAR, a Virtual TABULAR is created for each distinct graph that matches the pattern queried against Neo4j. The Virtual TABULAR consists of a table with the same name as the label shared by vertices. A default column 'ID' (of type string) is created to store each vertex of same label (i.e., rows). A new row is inserted for each vertex with different edge name into the corresponding table; For each distinct property of the vertex, an additional column is created typed according to the property extracted datatype. The cell information consists of the values extracted from the vertex's properties. There will be no edges between distinct graph pattern due to the results returned by the graph property.

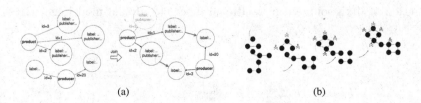

(a)   (b)

Fig. 3. (a) Join of temporary virtual GRAPHs (b) Multi-join algorithm of GRAPHs

2.4   Distributed Query Processor

Distributed Query Processor is the environment where queries are executed. OPTIMA calls for Graphx and Apache Spark to use two different virtual data models GRAPH and TABULAR, respectively. If deep learning predicts that the optimal virtual model is of type GRAPH, then for each relevant entity, one virtual GRAPH model is generated by wrappers. The wrappers use API to access data source and perform transformation. OPTIMA joins those GRAPHs into a

Final Virtual GRAPH (see Fig. 3a) using "multi-join algorithm" (see Fig. 3b) or TABULARs into a Final Virtual TABULAR using "incremental join algorithm". This is by mean of connections between star-shaped queries. However, GRAPH and TABULAR have different structures, for example, the interaction with GRAPH is possible by means of Graph Pattern Matching operations (Cypher-like), while the interaction with TABULAR is possible by SQL-like functions. SPARQL and star-shaped queries operations (e.g., limit) are translated into Virtual Data model operations (e.g., "take" in case of Graphx).

3 Experimental Setup

We conducted an experimental study to evaluate OPTIMA performance compared to the state-of-the-art SPARK-based Sequerall which uses dataframes (i.e. TABULAR) as virtual data model. We used five tables, to enable up to 4-chain joins. These tables are loaded in five different data sources Cassandra, MongoDB, CSV, Neo4j and MySQL. Table 1 shows the described information about data. We generated 5150 queries with 0–4 joins,0–45 selection, 0–16 filter, limit, OrderBy. We take 4120 queries for training the model and 1030 queries for validation. We run the evaluation on Ubuntu 64-bit with an Intel(R) Core(TM) i7-8550U CPU @ 1.80 GHz, allocating 8 GB of RAM. We measure the time taken by both systems from query submission to the delivery of the answer.

Table 1. Operations involved in queries.

	Q1	Q2	Q3	Q4	Q5	Q6	Q7	Q8	Q9	Q10	Q11	Q12	Q13	Q14	Q15	Q16	Q17	Q18	Q19	Q20
PROJECT	✓16	✓5	✓29	✓45	✓24	✓45	✓38	✓38	✓24	✓34	✓4	✓6	✓32	✓34	✓4	✓5	✓9	✓45	✓45	✓5
FILTER	✓16		✓12	✓1			✓5	✓1	✓1			~	✓1	✓1	✓4			✓2	✓3	
ORDERBY	✓1	✓1			✓1		✓1		✓1	✓1			✓1	✓1			✓1			
LIMIT	✓300	✓2		✓20	✓4	✓20	✓20	✓80		✓10			✓13	✓19	✓1000	✓1000				
DISTINCT	✓	✓	✓	✓	✓	✓	✓	✓	✓	✓	✓	✓	✓	✓	✓	✓	✓	✓	✓	✓

Table 2. Time per query of OPTIMA & Squerall

Application	Q1	Q2	Q3	Q4	Q5	Q6	Q7	Q8	Q9	Q10	Q11	Q12	Q13	Q14	Q15	Q16	Q17	Q18	Q19	Q20
OPTIMA	1291	1254	730	10299	10199	1553	7104	8442	10094	4694	2575	233	4673	4487	2397	2881	1698	4607	2804	5648
Squerall	4098	2519	3091	10283	10191	7984	7089	8427	10088	4684	2561	1400	4644	4469	3885	2875	3314	8742	9059	7407
Time difference	2807	1265	2361	16	8	6431	15	15	6	10	14	1167	29	18	1488	6	1616	4135	6255	1759

Table 2 illustrates the execution time returned by the two aforementioned applications. As can be observed, OPTIMA excels Squerall for queries that involve multiple joins. The time difference ranges from 0 to 80000 millisecond (ms). This difference is due to the predicted virtual data model e.g. Q19, Q20 in which the machine learning predicted that the Virtual model of type GRAPH is optimal. We observe also small difference in the execution time (ranging from 0 to 30 ms) in favor of Squerall compared to OPTIMA for queries that involve multiple projections e.g. Q7, Q10. This is explained by the fact that the optimal virtual model is identical to Squerall's, and both Squerall and OPTIMA used the same APIs to call data (wrapper), however, the data model prediction time added to OPTIMA makes it slightly slower than Squerall. Furthermore, the average execution time of Squerall is greater than 4000 ms compared to the average execution time of OPTIMA 2400 ms.

To check if the machine learning is reducing the overall execution time of OPTIMA by selecting the optimal virtual data model. We illustrate first the time taken by OPTIMA's components: machine learning algorithm, query execution over GRAPH model and query execution over TABULAR against SPARK-based Squerall. We run OPTIMA and Squerall over 1030 query. The average execution time of machine learning component is very short 12 ms while the average time for GRAPH is 1320 ms and TABULAR is 2862 ms. Results show that for the most queries the GRAPH is faster than TABULAR even with predictions time.

4   Related Work

Our study's scope focuses on works that query large-scale data sources using OBDA. Optique [2] is an OBDA platform that accesses both static and streaming data. It implements relational model (implicitly a TABULAR) as virtual model while querying data sources such as SQL databases. Ontario [1] focuses on the query rewriting, and federation, with a strong stress on RDF data as input. The virtual model used by Ontario is GRAPH model (explicitly an RDF). Squerall [3] a recent and close work to OPTIMA leverages Big Data engines Apache Spark and Presto to query on the fly large scale data sources. The virtual data model imposed by Presto is TABULAR, while Apache Spark uses dataframe as virtual model which is TABULAR. There is no work that (1) implements different virtual models (2) selects optimal one based on query behavior.

5   Conclusion

We implemented OPTIMA - an ontology-based big data access system that reduces query time execution by predicting the optimal virtual data model, GRAPH or TABULAR based on query behavior. The effective deep learning model built on top of OPTIMA's architecture extracts significant features such as the query plan, operations and predicts the optimal virtual data model that has the lowest query execution time. Experiment showed a reduction in query execution time of over 40% for the TABULAR model and over 30% for the GRAPH model selection.

Acknowledgments. The author Abderrahmane Khiat acknowledges the financial support of Fraunhofer Cluster of Excellence (CCIT) and Fraunhofer Research Center for Machine Learning (FZML).

References

1. Endris, K.M., Rohde, P.D., Vidal, M.-E., Auer, S.: Ontario: federated query processing against a semantic data lake. In: Hartmann, S., Küng, J., Chakravarthy, S., Anderst-Kotsis, G., Tjoa, A.M., Khalil, I. (eds.) DEXA 2019. LNCS, vol. 11706, pp. 379–395. Springer, Cham (2019). https://doi.org/10.1007/978-3-030-27615-7_29
2. Giese, M., et al.: Optique: zooming in on big data. Computer **48**(3), 60–67 (2015)
3. Mami, M.N., Graux, D., Scerri, S., Jabeen, H., Auer, S., Lehmann, J.: Squerall: virtual ontology-based access to heterogeneous and large data sources. In: Ghidini, C., et al. (eds.) ISWC 2019. LNCS, vol. 11779, pp. 229–245. Springer, Cham (2019). https://doi.org/10.1007/978-3-030-30796-7_15
4. Miloslavskaya, N., Tolstoy, A.: Big data, fast data and data lake concepts. Procedia Comput. Sci. **88**, 300–305 (2016)
5. Poggi, A., Lembo, D., Calvanese, D., De Giacomo, G., Lenzerini, M., Rosati, R.: Linking data to ontologies. In: Spaccapietra, S. (ed.) J. Data Semant. X. LNCS, vol. 4900, pp. 133–173. Springer, Heidelberg (2008). https://doi.org/10.1007/978-3-540-77688-8_5

Pattern Discovery

Q-VIPER: Quantitative Vertical Bitwise Algorithm to Mine Frequent Patterns

Thomas J. Czubryt, Carson K. Leung$^{(\boxtimes)}$ ⓘ, and Adam G. M. Pazdor

University of Manitoba, Winnipeg, MB, Canada
Carson.Leung@UManitoba.ca

Abstract. As a popular big data analytics and knowledge discovery task, frequent pattern mining aims to discover frequently occurring sets of items (e.g., merchandise items, events) from big data. Frequent patterns can be discovered horizontally by transaction-centric mining algorithms or vertically by item-centric mining algorithms. Regardless of their mining direction (horizontal or vertical), traditional frequent pattern mining algorithms aim to discover Boolean frequent patterns in the sense that patterns capture the presence (or absence) of items within the discovered patterns. However, there are many real-life situations, in which quantities of items within the patterns are important. For example, the quantity of items may also affect profits of selling the items within the discovered patterns. Hence, in this paper, we present a quantitative vertical bitwise algorithm to mine frequent patterns. This Q-VIPER algorithm first represents the big data as a collection of bitmaps. Each item-centric bitmap captures the presence or absence of a transaction containing the item, as well as the quantity of that item in each transaction. With this representation, our algorithm then vertically mines quantitative frequent patterns. When compared the existing MQA-M algorithm (which was built for quantitative frequent pattern mining), evaluation results show that our quantitative vertical bitwise Q-VIPER algorithm takes shorter runtime to mine quantitative frequent patterns.

Keywords: Frequent pattern mining · Quantitative data mining · Vertical pattern mining · Bitmap

1 Introduction

In the current era, big data [1] are everywhere. With advances in technology, high volumes of a wide variety of data (which may be of different levels of varsity) are generated and collected at a high velocity for numerous real-life applications and services (e.g., healthcare informatics [2], transportation analytics [3–5], business analytics [6, 7], social network analysis [8–10]). Embedded in these big data is implicit, previously unknown and potentially useful information and knowledge. This calls for big data management [11–14], as well as big data analytics and knowledge discovery [15, 16].

As a popular big data analytics and knowledge discovery task, association rule mining aims to discover rules that reveal interesting associations among the antecedents and

© The Author(s), under exclusive license to Springer Nature Switzerland AG 2022
R. Wrembel et al. (Eds.): DaWaK 2022, LNCS 13428, pp. 219–233, 2022.
https://doi.org/10.1007/978-3-031-12670-3_19

consequents of the rules. Generally, these rules are mined by first discovering frequent patterns and then using these discovered frequent patterns to form the rules. *Frequent pattern mining* [17–20] aims to discover frequently occurring sets of items (e.g., merchandise items, events) from big data. Given a series of transactions containing a set of items, frequent pattern mining seeks to determine the sets of items, which occur in a large number of transactions. In addition, we wish to discover interesting association rules. Association rules state that whenever a certain set of items occurs in a transaction, another set of items tends to occur in that transaction. The problems of frequent pattern mining and association rule mining form the basis of many real-life applications such as marketing in business, discovering biological patterns, studying human populations, and web log mining. Frequent pattern mining has been extended to the mining of other patterns such as network and graph mining [21–23], stream mining [24–26], uncertain pattern mining [27–29], and utility pattern mining [30–32].

Frequent patterns can be discovered horizontally by transaction-centric mining algorithms or vertically by item-centric mining algorithms [33–35]. The Apriori algorithm [36, 37] is an example of horizontal transaction-centric frequent pattern mining algorithms, with which data are represented as a collection of transactions. Each transaction captures the presence or absence of items. Alternatively, frequent patterns can also be discovered vertically. The VIPER (Vertical Itemset Partitioning for Efficient Rule-extraction) algorithm [38] is an example of vertical item-centric frequent pattern mining algorithms, with which data are represented as a collection of bitmaps. Each bitmap for an item captures which transactions contain the specific item. A bit of 1 in the i-th position indicates the presence of the item in the i-th transaction, whereas a bit of 0 in the i-th position indicates the absence of the item from the i-th transaction. An advantage of such a bitmap representation is that the size of bitmap collection is independent of the density of the data. Dense data contain more 1 s than 0 s, and vice versa for sparse data. The algorithm was shown to be efficient as it takes advantage of bitwise operations in the mining process.

Regardless of their mining direction (horizontal or vertical), traditional frequent pattern mining algorithms aim to discover Boolean frequent patterns in the sense that patterns capture the presence (or absence) of items within the discovered patterns. While traditional frequent pattern mining and association rule mining are useful in many contexts, they have a major limitation. This limitation is that in traditional frequent pattern mining, we assume that every transaction either contains an item or does not contain the item. In other words, an item is contained in a transaction 0 or 1 times. For this reason, we can also refer to traditional frequent pattern mining as *Boolean* frequent pattern mining. However, in many real-world scenarios, a transaction can contain an item more than one time. For example, a person at a grocery store may buy multiple apples. To address this shortcoming, the notion of quantitative association rule mining or *quantitative* frequent pattern mining [39, 40] was studied. Quantitative frequent pattern mining is essentially an extension of frequent pattern mining to allow transactions to contain an item more than once. Rather than just trying to find items (which commonly occur in transactions), there is a demand for discovering commonly occurring quantities of items. For example, in Boolean frequent itemset mining, we may discover that bananas are a frequently purchased item. In quantitative frequent pattern mining, we may discover that customers frequently purchase at least five bananas at a time. As another example, the quantity of items may also affect profits of selling the items within the discovered patterns.

By discovering quantitative frequent patterns and quantitative association rules, we can obtain more interesting results than we would if Boolean association rule mining were used instead. In addition to receiving information about which items commonly occur together in transactions, we also obtain information regarding how many of each of those items tend to occur in transactions. MQA-M algorithm [39] extends the Apriori algorithm to mine *quantitative frequent patterns* (aka *itemexpsets*) horizontally.

In this paper, we present a vertical bitwise algorithm to mine quantitative frequent patterns (i.e., itemexpsets) *vertically*. The resulting Q-VIPER algorithm first represents the big data as a col-lection of bitmaps. Each item-centric bitmap captures the presence or absence of a transaction containing the item, as well as the quantity of that item in each transaction. With this representation, our algorithm then vertically mines quantitative frequent patterns. When compared the existing MQA-M algorithm (which was built for quantitative frequent pattern mining), evaluation results show that our quantitative vertical bitwise Q-VIPER algorithm requires shorter runtime to mine frequent patterns. Our *key contributions* in this paper include our Q-VIPER algorithm and its pruning rules.

The remainder of this paper is organized as follows. We begin by presenting the mathematical framework for quantitative frequent pattern mining in Sect. 2. We discuss previously used algorithms of interest, such as the Apriori, MQA-M, and VIPER algorithms. Then, we formally introduce our Q-VIPER algorithms in Sect. 3. Pseudo-code and an example are provided for the algorithm. Section 4 contains analysis of the algorithm and evaluation to compare our Q-VIPER with related works. Finally, we conclude in Sect. 5.

2   Background and Related Works

Before describing our algorithm for quantitative frequent pattern mining, we formally define quantitative frequent patterns and review relevant algorithms.

2.1   Vertical Boolean Frequent Pattern Mining with the VIPER Algorithm

Recall from Sect. 1 that the **VIPER** (Vertical Itemset Partitioning for Efficient Rule-extraction) algorithm [38] is an example of vertical item-centric frequent pattern mining algorithms, with which data are represented as a collection of bitmaps. Each bitmap for an item captures which transactions contain the specific item. A bit of 1 in the i-th position indicates the presence of the item in the i-th transaction, whereas a bit of 0 in the i-th position indicates the absence of the item from the i-th transaction.

Let us discuss the difference between the horizontal transaction database and the vertical transaction database. Horizontal transaction databases refer to the standard representation of transactions, where a set of items is associated with each transaction [36, 37]. The Apriori algorithm uses the horizontal representation. On the other hand, one can represent the transaction database in a "vertical" format [38]. A bitmap for an item can represent a transaction database in a vertical format by putting a "1"-bit in the i-th position indicates the presence of the item in the i-th transaction and a "0"-bit the i-th position indicates the absence of the item from the i-th transaction. For example, for transactions $T_1 = \{a, b\}$ and $T_2 = \{b\}$, the corresponding vertical representation of the transaction database is $bitmap(a) = [10 \ldots]$ and $bitmap(b) = [11 \ldots]$. Each bitmap is of the same length, and its length equals to the number of transactions. The VIPER algorithm makes use of the vertical representation.

Like the Apriori algorithm, let C_k and L_k be the sets containing candidate itemsets and frequent itemsets respectively of size k. First, the VIPER algorithm determines which itemsets are in L_1. It then computes the support of any itemset simply by counting (or summing) the number of "1"-bits in its corresponding bitmap. Mathematically, for an itemset X, $\sup(X) = \sum_i bitmap(X, i)$, where bitmap(X, i) denotes the i-th position of bitmap of X. After computing the support for every item occurring in the transaction database, L_1 contains each item with a support greater than or equal to *minsup*.

After determining L_1, the main loop of the VIPER algorithm is be executed. This loop is very similar in structure to the loop in the Apriori algorithm. Consider the first loop iteration with $k = 2$. The first part of the loop involves generating C_k from L_{k-1}. This uses the same candidate generation method in the Apriori algorithm (i.e., performing a self-join on L_{k-1} and pruning the resulting set). Next, it forms bitmap corresponding to each itemset in C_k. Suppose that, for some itemset $X \in C_k$, W is an itemset containing the first $(k-2)$ items in X, y is the second last item in X, and z is the last item in X. Then, $X = W \cup \{y\} \cup \{z\}$. The algorithm computes the bitmap of X as the cross-product of $(W \cup \{y\})$ and $(W \cup \{z\})$, i.e., bitmap(X, i) = bitmap(W $\cup \{y\}, i$) × bitmap(W $\cup \{z\}, i$)}. Next, it computes the support of each pattern in C_k by summing the number of "1"-bits in the resulting bitmap(X). The frequent patterns in L_k are computed as the candidate patterns in C_k with a support that is at least *minsup*. At the end of a loop iteration, increase k by 1 and continue iterating through the main loop (if necessary). The loop stops iterating when L_{k-1} is empty. In a similar fashion to the Apriori algorithm, all the frequent patterns ($\bigcup_k L_k$) are returned by the algorithm.

2.2 Quantitative Association Rule Mining

For quantitative association rule mining [39, 40], suppose that $I = \{i_1, i_2, \cdots, i_m\}$ is the set of all items that can be found in a transaction database for some positive integer m,. Then, a transaction can be represented as $T = \{(e_1, f_1), (e_2, f_2), \cdots, (e_t, f_t)\}$ for some positive integer t, where each $e_i \in I$, such that $e_i \neq e_j$ whenever $i \neq j$, and each f_i is a positive integer. The quantitative transaction database is $D = (T_1, T_2, \cdots, T_n)$, which is the set of all transactions. Each transaction has a unique numeric identifier TID.

An ***itemexp*** (short for ***item-expression***) is an ordered triplet of the form $(p, \otimes, q)$, where $p \in I$, $\otimes \in \{=, \geq, \leq\}$, and q is a positive integer. We represent an itemexp as $(p \otimes q)$. Then, an ***itemexpset*** can be defined as a set $X = \{x_1, x_2, \cdots, x_k\}$ for some positive integer k, where each $x_i = (p_i, \otimes_i, q_i)$ is an itemexp such that $p_i \neq p_j$ whenever $i \neq j$.

For a transaction $T = \{(e_1, f_1), (e_2, f_2), \cdots, (e_t, f_t)\}$ and an itemexpset $X = \{x_1, x_2, \cdots, x_k\}$ with $x_i = (p_i, \otimes_i, q_i)$, T ***satisfies*** X if for every $i \in \{1, 2, \cdots, k\}$, there exists some $j \in \{1, 2, \cdots, t\}$ such that $p_i = e_j$ and the expression $(f_j \otimes_i q_i)$ is true. For example, if $T = \{(a, 2), (b, 3), (c, 1)\}$ and $X = \{(a = 2), (b \geq 1)\}$, then T satisfies

X. However, if $T = \{(a, 2), (b, 3), (c, 1)\}$ and $X = \{(a \le 2), (c \ge 2)\}$, then T does not satisfy X because X requires $c \ge 2$ but c only occurs once in T.

If an itemexpset X contains an itemexp of the form $(p \le q)$ where $p \in I$ and q is a positive integer, then for a transaction T to satisfy X, the item p must still occur in T at least once, even though $0 < q$. In other words, the number of occurrences of item p in T must be in the interval $[1, q]$. For example, if $T = \{(a, 1)\}$ and $X = \{(b \le 2)\}$, then T does not satisfy X, even though the number of occurrences of b in T is at most 2. By including this restriction, many itemexpsets are prevented from being considered where an item can occur zero times.

For an itemexpset X, the **support** of X (denoted $\sup(X)$) is defined as the number of transactions in D that satisfy X. Now, let *minsup* be some non-negative real number. Then, X is a frequent (large) itemexpset if $\sup(X) \ge$ *minsup*.

Just as association rules can be defined for Boolean frequent pattern mining, they can also be defined for quantitative frequent pattern mining. For two itemexpsets X and Y, the association rule $X \Rightarrow Y$ is **interesting** if:

- there are no common items between X and Y,
- $\sup(X \cup Y) \ge$ *minsup*, and
- for some confidence value *minconf* $\in [0, 1]$, $\frac{\sup(X \cup Y)}{\sup(X)} \ge$ *minconf*.

For example, it is possible for $\{(a \ge 2)\} \Rightarrow \{(b \le 3), (c = 1)\}$ to be an interesting association rule if the support and confidence values are satisfied. However, $\{(a = 5)\} \Rightarrow \{(a \ge 3)\}$ cannot be an interesting association rule, since the item a appears on both sides of the rule.

2.3 Horizontal Quantitative Frequent Pattern Mining with the MQA-M Algorithm

The **MQA-M** (Mining Quantitative Association rules with Multiple comparison operators) [39] is an algorithm for mining quantitative frequent patterns. The MQA-M algorithm is very similar to the Apriori algorithm except that it is generalized to handle quantitative transaction databases.

For any positive integer k, let C_k be the set of candidate itemexpsets containing k itemexps and let L_k be the set of candidate itemexpsets containing k itemexps. Like in the Apriori algorithm, $L_k \subseteq C_k$.

The MQA-M algorithm starts by generating C_1. Suppose that *item_max*$[p]$ represents the maximum number of times an item p appears in a transaction. For example, if the quantitative database consists of transactions $T_1 = \{(a, 1)\}$ and $T_2 = \{(a, 3)\}$, then *item_max*$[a] = 3$ because the highest number of times a appears in a transaction is 3. Then, for each item p appearing in the quantitative transaction database, add every itemexpset of the form $\{(p, \otimes, q)\}$ to C_1, where $\otimes \in \{=, \ge, \le\}$ and $q \in \{1, \cdots, \text{*item_max*}[p]\}$. The algorithm computes the support of each itemexpset in C_1 by iterating through the transactions and checking each itemexpset in C_1 to see if it should increment the support of that itemexpset. It increments the support of an itemexpset if the transaction satisfies that itemexpset. Let $k = 1$. Then, L_1 becomes the set of all itemexpsets with a support that is at least *minsup*. The algorithm removes some itemexpsets from L_1 using two pruning rules [39]:

- Suppose that X contains an itemexp of the form $(z \leq r)$, where z is an item and r is a positive integer. The first pruning rule states that if there is another itemexpset Y in L_k with the same support as X which is the same as X except that $(z \leq r)$ is replaced by $(z \leq r + 1)$, then Y can be pruned from L_k.
- Suppose that X contains an itemexp of the form $(z \geq r)$, where z is an item and r is a positive integer. The second pruning rule states that if there is another itemexpset Y in L_k with the same support as X which is the same as X except that $(z \geq r)$ is replaced by $(z \geq r - 1)$, then Y can be pruned from L_k.

Similar to the Apriori algorithm, the MQA-M also has a main loop. It first runs the loop with $k = 2$. The loop body begins with generating C_k from L_{k-1}. C_k is initially generated using a self join on L_{k-1}, like in the Apriori algorithm. If 2 itemexpsets in L_{k-1} have the same first $(k - 2)$ itemexps, then it generates an itemexpset in C_k consisting of those $(k - 2)$ itemexps and the last itemexp in the 2 itemexpsets in L_{k-1}. However, it imposes an additional restriction that it does not create an itemexpset in C_k where there are 2 itemexps referring to the same item. For example, if L_1 contains $\{(a = 1)\}$ and $\{(a \geq 2)\}$, it does not form $\{(a = 1), (a \geq 2)\}$ in C_2. After the join step, it prunes itemexpsets from C_k with a subset containing $(k - 1)$ itemexps where that subset is not in L_{k-1}. It gets L_k from C_k using the same procedure that was used to obtain L_1. Using the two aforementioned pruning rules, it removes some itemsets from L_k. At the end of the loop body, it increments k and repeats the previous steps (if necessary). The loop terminates when L_{k-1}. is empty. Afterwards, it returns $\bigcup_k L_k$, which contains all the interesting frequent itemexpsets.

3 Vertical Quantitative Frequent Pattern Mining with Our Q-VIPER Algorithm

3.1 Vertical Representation of Quantitative Data

To represent quantitative transaction databases in a vertical format, for each item that occurs in the transaction database, we store it as a set of pairs. Each pair contains a transaction ID associated with that item and the number of occurrences of the item in the transaction. Since we are storing a pair, we can call these sets "pairsets". For example, if we have the transactions $T_1 = \{(a, 1)\}$ and $T_2 = \{(a, 3)\}$, then the transaction database can be represented vertically using *pairset*$(a) = \{(T_1, 1), (T_2, 3)\}$. It is useful to convert the quantitative transaction database to this vertical format when implementing the Q-VIPER algorithm.

We can also define *bitmaps* for quantitative association rule mining. For any itemexpset X, *bitmap*(X) is defined as the set of transaction IDs corresponding to transactions which satisfy X. Going back to the example from the previous paragraph, we can determine that *bitmap*$(\{(a \geq 1)\}) = [11\ldots]$ while *bitmap*$(\{(a = 3)\}) = [01\ldots]$. When X is an itemexpset containing at least 2 itemexps, we can break down X as $X = W \cup \{y\} \cup \{z\}$, where W is an itemexpset with 2 fewer elements than X and y and z are itemexps. Like bitmap for Boolean frequent pattern mining, we have:

- a recursive equation to compute the cross-product to indicate the presence (or absence of transactions that contain the specific item, and
- another recursive equation to compute the minimum to indicate the quantity of the specific item.

We use this recursive definition to generate bitmaps for itemexpsets containing at least two "1"-bits when running the Q-VIPER algorithm. The support of an itemexpset X can be computed simply by counting or summing the number of "1"-bits in its bitmap.

3.2 Q-VIPER Algorithm

Here, let us describe how **our Q-VIPER algorithm** discovers quantitative frequent patterns vertically. For any integer $k \geq 1$, define C_k to be the set of candidate itemexpsets with size k and L_k to be the set of frequent itemexpsets with size k. First, we convert the quantitative transaction database into a vertical format if it is not already. The vertical format is useful for computing the bitmaps corresponding to the itemexpsets in C_1. The next step of our algorithm is to compute all itemexpsets in C_1. Each of those itemexpsets consists of a single itemexp of the form (*item, operation, quantity*), where *item* is an item in the transaction database, *operation* $\in \{=, \geq, \leq\}$, and *quantity* $\in \{1, \cdots, item\_max[item]\}$. We compute *item_max[item]* as the maximum number of times *item* appears in a transaction, over all transactions in the transaction database. After computing C_1, we compute the bitmap associated with each itemexpset in C_1. The bitmaps can easily be computed from the vertical representation of the quantitative transaction database. We then compute the support of each itemexpset in C_1 by counting or summing the number of "1"-bits in its corresponding bitmaps. The itemexpsets in L_1 are itemexpsets in C_1 with a support $\geq$ *minsup*. Finally, we remove some itemexpsets from L_1 based on our two *new pruning rules*, which will be described in Sect. 3.3.

Next, we set $k = 2$ and begin executing the main loop. The first step in the main loop body is to generate C_k using L_{k-1}. This can be done using the same method that was used in the MQA-M algorithm. We initially create C_k by performing a self join on L_{k-1}. If there are 2 itemexpsets in L_{k-1} where the first $(k-2)$ itemexps in those itemexpsets are the same and the last itemexp in those itemexpsets refer to different items, then we add an itemexpset to C_k consisting of those first $(k-2)$ itemexps as well as the last itemexp of both itemexpsets. Afterwards, we prune any itemexpset in C_k that contains a sub-itemexpset with $(k-1)$ itemexps that is not in L_{k-1}. The next step is to create bitmaps corresponding to every itemexpset in C_k. This can be done using the recursive definition for bitmaps, which was discussed earlier. After computing the bitmaps, we can easily compute the support of each itemexpset in C_k. Any itemexpset in C_k with a support $\geq$ *minsup* is added to L_k. Using the two pruning rules, we remove some uninteresting itemexpsets from L_k, if necessary. After pruning L_k, we have reached the end of the loop body, so we increment k and repeat the main steps again if necessary. The main loop stops running once L_k is empty. Our Q-VIPER algorithm returns $\bigcup_k L_k$, which contains all interesting frequent itemexpsets.

3.3 Our New Pruning Rules for Q-VIPER Algorithm

As mentioned above, there are two pruning rules which we use to remove unnecessary itemexpsets from L_k for each integer $k \geq 1$. The pruning rules we use are more general than the ones described in Sect. 2.3 (for the MQA-M algorithm). Our improved pruning rules remove some uninteresting itemexpsets, which were not removed in the original pruning rules. Assume that X is an itemexpset in L_k. Then, the pruning rules are described as follows:

1. Suppose that X contains an itemexp of the form $(z \leq r)$, where z is an item and r is a positive integer. The first pruning rule states that if there is another itemexpset Y in L_k with the same support as X which is the same as X except that $(z \leq r)$ is replaced by $(z \leq r + s)$ for some positive integer s, then Y can be pruned from L_k.
2. Suppose that X contains an itemexp of the form $(z \geq r)$, where z is an item and r is a positive integer. The second pruning rule states that if there is another itemexpset Y in L_k with the same support as X which is the same as X except that $(z \geq r)$ is replaced by $(z \geq r - s)$ for some positive integer s, then Y can be pruned from L_k.

The difference between the original pruning rules and our new pruning rules is that the new pruning rules can handle differences in quantity greater than 1. Instead of considering itemexpsets of the form $(z \leq r + 1)$ or $(z \geq r - 1)$, we consider the more general cases of $(z \leq r + s)$ or $(z \geq r - s)$ for some positive integer s. As a result, these rules eliminate at least as many itemexpsets from L_k as the original pruning rules.

Example 1. Suppose that L_2 contains $\{(a = 1), (b \geq 3)\}$ and $\{(a = 1), (b \geq 6)\}$ before pruning and that those itemexpsets have the same support. Using the original pruning rules that were used in MQA-M, neither itemexpset would be pruned. However, using the new pruning rules, we would prune $\{(a = 1), (b \geq 3)\}$ from L_2. ∎

As observed, our improved pruning rules are more powerful in removing redundant frequent itemexpsets. See Fig. 1 for pseudo code of the resulting Q-VIPER algorithm.

Example 2. Suppose we set *minsup* $= 2$ and have 3 transactions in a quantitative transaction database: T1 $\{a:2\}$, T2 $\{a:3, b:1\}$, and T3 $\{b:1\}$.

To generate C_1, we note that *item_max*$[a] = 3$ and *item_max*$[b] = 1$, since the highest number of occurrences of a in a transaction is 3 and the highest number of occurrences of b in a transaction is 1. For generating C_1, we must generate every combination of an item, comparison operation, and quantity. This gives us a total of 12 candidate itemexpsets. For each itemexpset in C_1, we compute its corresponding bitmap. The support of those itemexpsets is equal to the number of "1"-bits in the bitmap. We present the itemexpsets X in C_1, their bitmaps, and their supports in the first three column of Table 1.

Then, we can obtain L_1 from this by only keeping the itemexpsets in C_1 with a support $\geq 2 = minsup$. Therefore, we initially get L_1 with six itemexpsets (before pruning)—as shown in the fourth and fifth columns—in Table 1.

Q-VIPER algorithm (quantitative transaction database TDB, *minsup* threshold)
  if (TDB is not in vertical format)
  then convert TDB to vertical format

$C_1 = \varnothing$
for each item in TDB do
  Item_max[item] = max #items in a transaction
  for each quantity in {1, ..., item_max[item]} do
    add {item, operator, quantity} to C_1

  Bitmap[1] = createBitmap1 (TDB, C_1)
  computeSupport (C_1, Bitmap[1])
  $L_1 = \{c \in C_1 \mid \text{sup}(c) \geq minsup\}$
  applyOurPruningRules (L_1)

  for (k=2; $L_{k-1} \neq \varnothing$; k++) do
    Ck = generateCandidate (L_{k-1})
    Bitmap[k] = computeBitmap(C_k, Bitmap[k–1])
    computeSupport (C_k, Bitmap[k])
    $L_k = \{c \in C_k \mid \text{sup}(c) \geq minsup\}$
    applyOurPruningRules (L_k)

  return $\bigcup_k L_k$

Fig. 1. Pseudo code of our Q-VIPER algorithm.

By our Pruning Rule 2, since both $\{(a \geq 1)\}$ and $\{(a \geq 2)\}$ are in L_1 and have the same support, the itemexpset $\{(a \geq 1)\}$ is no longer interesting and can be pruned. Therefore, L_1 ends up with the only five itemexpsets, as shown in the last two columns of Table 1.

Afterwards, the main loop is executed with $k = 2$. We begin with the generation of C_2. The first step in generating C_2 is to do a self join on L_1. In this scenario, this means getting pairs of itemexps where the itemexps refer to different items. This yields 6 different itemexpsets in C_2. None of those itemexpsets is pruned because for all itemexpsets in C_2, there are no sub-itemexpsets which are not in L_1. Computing the cross-product, we then obtain the bitmap associated with each itemexpset in C_2. Finally, the support for those itemexpsets is computed by counting or summing the number of "1"-bits in the bitmaps. The itemexpsets in C_2, their bitmaps, and their supports are shown in Table 2.

Table 1. Candidate and frequent itemexpsets of size 1.

C_1			L_1 before pruning		L_1 after pruning	
X	bitmap(X)	sup(X)	X	sup(X)	X	sup(X)
$\{(a = 1)\}$	[000]	0				
$\{(a = 2)\}$	[100]	1				
$\{(a = 3)\}$	[010]	1				
$\{(a \geq 1)\}$	[110]	2	$\{(a \geq 1)\}$	2		
$\{(a \geq 2)\}$	[110]	2	$\{(a \geq 2)\}$	2	$\{(a \geq 2)\}$	2
$\{(a \geq 3)\}$	[010]	1				
$\{(a \leq 1)\}$	[000]	0				
$\{(a \leq 2)\}$	[100]	1				
$\{(a \leq 3)\}$	[110]	2	$\{(a \leq 3)\}$	2	$\{(a \leq 3)\}$	2
$\{(b = 1)\}$	[011]	2	$\{(b = 1)\}$	2	$\{(b = 1)\}$	2
$\{(b \geq 1)\}$	[011]	2	$\{(b \geq 1)\}$	2	$\{(b \geq 1)\}$	2
$\{(b \leq 1)\}$	[011]	2	$\{(b \leq 1)\}$	2	$\{(b \leq 1)\}$	2

Table 2. Candidate itemexpsets of size 2.

C_2		
X	bitmap(X)	sup(X)
$\{(a \geq 2), (b = 1)\}$	$[110] \times [011]^T = [010]$	1
$\{(a \geq 2), (b \geq 1)\}$	$[110] \times [011]^T = [010]$	1
$\{(a \geq 2), (b \leq 1)\}$	$[110] \times [011]^T = [010]$	1
$\{(a \leq 3), (b = 1)\}$	$[110] \times [011]^T = [010]$	1
$\{(a \leq 3), (b \geq 1)\}$	$[110] \times [011]^T = [010]$	1
$\{(a \leq 3), (b \leq 1)\}$	$[110] \times [011]^T = [010]$	1

As all itemexpsets in C_2 have a support < 2, $L_2 = \emptyset$. Therefore, the pruning rules do not remove any itemexpsets. Since L_2 is empty, the loop does not execute for $k = 3$. Thus, the frequent itemexpsets returned by the algorithm only consists of the six itemexpsets in L_1. ∎

4 Evaluation

To evaluate our Q-VIPER algorithm, we compared it with the existing MQA-M algorithm [39]. The performance of the algorithms is assessed using four different quantitative transaction databases:

- two synthetic dataset: Here, we assume that there are n transactions and $|I|$ different items. Each item has a probability *prob* of occurring in a particular transaction, where $0 \leq prob \leq 1$. If the item appears in the transaction, then the number of occurrences of that item follows a *Poisson*(λ) distribution plus 1. We set $n = 1000$, $|I| = 50$, and $\lambda = 1$. The values of *prob* for these two quantitative transaction databases are 0.2 and 0.8. These quantitative transaction databases considered as sparse and dense, respectively, i.e.,

1 **sparse synthetic dataset**, with *prob*=0.2; and
2 **dense synthetic dataset**, with *prob*=0.8.

- two real-life datasets from UCI ML Repository [41]: Here, we modified the chess and mushroom datasets to make them quantitative transaction databases. Whenever an item occurs in a transaction, instead of it only occurring once, its number of occurrences follows a *Poisson*($\lambda = 1$) distribution plus 1, i.e.,

1 modified **chess** dataset; and
2 modified **mushroom** dataset.

The two algorithms for quantitative frequent itemset mining (i.e., MQA-M [39] and our Q-VIPER) have been implemented in the Python language. The algorithms were run on a Windows 10 Nitro AN515–55 laptop using an Intel® Core™ i5-10300H CPU at 2.50 GHz and 8.00 GB RAM. To keep the comparisons between the three algorithms fair, many of the same functions are used between the algorithms. Procedures including generating candidate itemexpsets, determining frequent itemexpsets, and using the pruning rules on the frequent itemexpsets are kept the same among all three algorithms. When we implement the MQA-M algorithm, we use the improved pruning rules used in Q-VIPER rather than the pruning rules originally used with MQA-M. This allows the simulations to emphasize the differences between the algorithms.

We run the main code once for every quantitative transaction database. For each of those quantitative transaction databases, we use a sequence of *minsup* values. The sequence depends on the quantitative transaction database being used so that we see interesting results and that the algorithms do not take too long to run. For each combination of a quantitative transaction database and a value for *minsup*, the MQA-M, Q-VIPER algorithms are be run and timed.

Figure 2 shows the runtimes of each of the two algorithms for a variety of values of *minsup* for each of the four quantitative transaction databases. The runtime (in seconds) is shown on the y-axis while the value of *minsup* is given on the x-axis. In all cases, our Q-VIPER outperforms the existing MQA-M algorithm.

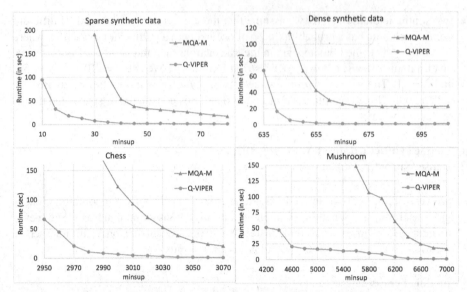

Fig. 2. The runtimes of the existing MQA-M algorithm and our Q-VIPER algorithms for quantitative frequent itemset mining for various *minsup* values.

5 Conclusions

In this paper, we presented our vertical quantitative frequent itemset mining called Q-VIPER. This Q-VIPER algorithm first represents the big data as a collection of bitmaps. Each item-centric bitmap captures the presence or absence of a transaction containing the item, as well as the quantity of that item in each transaction. With this representation, our algorithm then vertically mines quantitative frequent patterns. During the mining process, our new pruning rules reduce the mining space, and thus reduce the runtime. When compared the existing MQA-M algorithm (which was built for quantitative frequent pattern mining), evaluation results show that our quantitative vertical bitwise Q-VIPER algorithm requires shorter runtime to mine frequent patterns. As *ongoing and future work*, we explore ways to further enhance the mining of quantitative frequent patterns and to extend this work to mine other quantitative patterns.

Acknowledgement. This work is partially supported by Natural Sciences and Engineering Research Council of Canada (NSERC) and University of Manitoba.

References

1. Bemarisika, P., Totohasina, A.: ERAPN, an algorithm for extraction positive and negative association rules in big data. In: Ordonez, C., Bellatreche, L. (eds.) DaWaK 2018. LNCS, vol. 11031, pp. 329–344. Springer, Cham (2018). https://doi.org/10.1007/978-3-319-98539-8_25
2. Leung, C.K., Fung, D.L.X., Hoi, C.S.H.: Health analytics on COVID-19 data with few-shot learning. In: Golfarelli, M., Wrembel, R., Kotsis, G., Tjoa, A.M., Khalil, I. (eds.) DaWaK 2021. LNCS, vol. 12925, pp. 67–80. Springer, Cham (2021). https://doi.org/10.1007/978-3-030-86534-4_6
3. Audu, A.-R.A., Cuzzocrea, A., Leung, C.K., MacLeod, K.A., Ohin, N.I., Pulgar-Vidal, N.C.: An intelligent predictive analytics system for transportation analytics on open data towards the development of a smart city. In: Barolli, L., Hussain, F.K., Ikeda, M. (eds.) CISIS 2019. AISC, vol. 993, pp. 224–236. Springer, Cham (2020). https://doi.org/10.1007/978-3-030-22354-0_21
4. Leung, C.K., Braun, P., Hoi, C.S.H., Souza, J., Cuzzocrea, A.: Urban analytics of big transportation data for supporting smart cities. In: Ordonez, C., Song, I.-Y., Anderst-Kotsis, G., Tjoa, A.M., Khalil, I. (eds.) DaWaK 2019. LNCS, vol. 11708, pp. 24–33. Springer, Cham (2019). https://doi.org/10.1007/978-3-030-27520-4_3
5. Leung, C.K., Braun, P., Pazdor, A.G.M.: Effective classification of ground transportation modes for urban data mining in smart cities. In: Ordonez, C., Bellatreche, L. (eds.) DaWaK 2018. LNCS, vol. 11031, pp. 83–97. Springer, Cham (2018). https://doi.org/10.1007/978-3-319-98539-8_7
6. Ahn, S., et al.: A fuzzy logic based machine learning tool for supporting big data business analytics in complex artificial intelligence environments. In: FUZZ-IEEE, pp. 1259–1264 (2019)
7. Morris, K.J., et al.: Token-based adaptive time-series prediction by ensembling linear and non-linear estimators: a machine learning approach for predictive analytics on big stock data. In: IEEE ICMLA, pp. 1486–1491 (2018)
8. Braun, P., Cuzzocrea, A., Jiang, F., Leung, C.K.-S., Pazdor, A.G.M.: MapReduce-based complex big data analytics over uncertain and imprecise social networks. In: Bellatreche, L., Chakravarthy, S. (eds.) DaWaK 2017. LNCS, vol. 10440, pp. 130–145. Springer, Cham (2017). https://doi.org/10.1007/978-3-319-64283-3_10
9. Jiang, F., Leung, C.K.-S.: Mining interesting "following" patterns from social networks. In: Bellatreche, L., Mohania, M.K. (eds.) DaWaK 2014. LNCS, vol. 8646, pp. 308–319. Springer, Cham (2014). https://doi.org/10.1007/978-3-319-10160-6_28
10. Leung, C.K.: Mathematical model for propagation of influence in a social network. In: Alhajj, R., Rokne, J. (eds.) Encyclopedia of Social Network Analysis and Mining, 2nd edn., pp. 1261–1269. Springer, New York, NY (2018). https://doi.org/10.1007/978-1-4939-7131-2_110201
11. Leung, C.K., et al.: Parallel social network mining for interesting 'following' patterns. Concurr. Comput. Pract. Exp. **28**(15), 3994–4012 (2016)
12. Arora, N.R., Lee, W., Leung, C.K.-S., Kim, J., Kumar, H.: Efficient fuzzy ranking for keyword search on graphs. In: Liddle, S.W., Schewe, K.-D., Tjoa, A.M., Zhou, X. (eds.) DEXA 2012, Part I. LNCS, vol. 7446, pp. 502–510. Springer, Heidelberg (2012). https://doi.org/10.1007/978-3-642-32600-4_38
13. Eom, C.S., et al.: Effective privacy preserving data publishing by vectorization. Inf. Sci. **527**, 311–328 (2020)
14. Olawoyin, A.M., Leung, C.K., Choudhury, R.: Privacy-preserving spatio-temporal patient data publishing. In: Hartmann, S., Küng, J., Kotsis, G., Tjoa, A.M., Khalil, I. (eds.) DEXA 2020, Part II. LNCS, vol. 12392, pp. 407–416. Springer, Cham (2020). https://doi.org/10.1007/978-3-030-59051-2_28

15. Leung, C.K.-S., Jiang, F.: Big data analytics of social networks for the discovery of "following" patterns. In: Madria, S., Hara, T. (eds.) DaWaK 2015. LNCS, vol. 9263, pp. 123–135. Springer, Cham (2015). https://doi.org/10.1007/978-3-319-22729-0_10

16. Souza, J., Leung, C.K., Cuzzocrea, A.: An innovative big data predictive analytics framework over hybrid big data sources with an application for disease analytics. In: Barolli, L., Amato, F., Moscato, F., Enokido, T., Takizawa, M. (eds.) AINA 2020. AISC, vol. 1151, pp. 669–680. Springer, Cham (2020). https://doi.org/10.1007/978-3-030-44041-1_59

17. Couronne, C., Koptelov, M., Zimmermann, A.: PrePeP: a light-weight, extensible tool for predicting frequent hitters. In: Dong, Y., Ifrim, G., Mladenic, D., Saunders, C., Van Hoecke, S. (eds.) ECML PKDD 2020, Part V. Applied Data Science and Demo Track. LNCS, vol. 12461. Springer, Cham (2021). https://doi.org/10.1007/978-3-030-67670-4_41

18. Fischer, J., Vreeken, J.: Sets of robust rules, and how to find them. In: Brefeld, U., Fromont, E., Hotho, A., Knobbe, A., Maathuis, M., Robardet, C. (eds.) ECML PKDD 2019. LNCS, vol. 11906. Springer, Cham (2020). https://doi.org/10.1007/978-3-030-46150-8_3

19. Leung, C.K.-S., Tanbeer, S.K.: Mining popular patterns from transactional databases. In: Cuzzocrea, A., Dayal, U. (eds.) DaWaK 2012. LNCS, vol. 7448, pp. 291–302. Springer, Heidelberg (2012). https://doi.org/10.1007/978-3-642-32584-7_24

20. Seiffarth, F., Horvath, T., Wrobel, S.: Maximal closed set and half-space separations in finite closure systems. In: Brefeld, U., Fromont, E., Hotho, A., Knobbe, A., Maathuis, M., Robardet, C. (eds.) Machine ECML PKDD 2019, Part I. LNCS, vol. 11906. Springer, Cham (2020). https://doi.org/10.1007/978-3-030-46150-8_2

21. Alam, M.T., Ahmed, C.F., Samiullah, M., Leung, C.K.: Mining frequent patterns from hypergraph databases. In: Karlapalem, K., et al. (eds.) PAKDD 2021, Part II. LNCS (LNAI), vol. 12713, pp. 3–15. Springer, Cham (2021). https://doi.org/10.1007/978-3-030-75765-6_1

22. Chowdhury, M.E.S., et al.: A new approach for mining correlated frequent subgraphs. ACM Trans. Manag. Inf. Syst. **13**(1), 9:1–9:28 (2022)

23. Leung, C.K.-S., Carmichael, C.L., Teh, E.W.: Visual analytics of social networks: mining and visualizing co-authorship networks. In: Schmorrow, D.D., Fidopiastis, C.M. (eds.) FAC 2011. LNCS (LNAI), vol. 6780, pp. 335–345. Springer, Heidelberg (2011). https://doi.org/10.1007/978-3-642-21852-1_40

24. Cuzzocrea, A., Jiang, F., Leung, C.K., Liu, D., Peddle, A., Tanbeer, S.K.: Mining popular patterns: a novel mining problem and its application to static transactional databases and dynamic data streams. In: Hameurlain, A., Küng, J., Wagner, R., Cuzzocrea, A., Dayal, U. (eds) Transactions on Large-Scale Data- and Knowledge-Centered Systems XXI. LNCS, vol. 9260. Springer, Heidelberg (2015). https://doi.org/10.1007/978-3-662-47804-2_6

25. Jiang, F., Leung, C.K.-S.: Stream mining of frequent patterns from delayed batches of uncertain data. In: Bellatreche, L., Mohania, M.K. (eds.) DaWaK 2013. LNCS, vol. 8057, pp. 209–221. Springer, Heidelberg (2013). https://doi.org/10.1007/978-3-642-40131-2_18

26. Leung, C.K.-S., Jiang, F.: Frequent pattern mining from time-fading streams of uncertain data. In: Cuzzocrea, A., Dayal, U. (eds.) DaWaK 2011. LNCS, vol. 6862, pp. 252–264. Springer, Heidelberg (2011). https://doi.org/10.1007/978-3-642-23544-3_19

27. Leung, C.K.-S., MacKinnon, R.K.: Balancing tree size and accuracy in fast mining of uncertain frequent patterns. In: Madria, S., Hara, T. (eds.) DaWaK 2015. LNCS, vol. 9263, pp. 57–69. Springer, Cham (2015). https://doi.org/10.1007/978-3-319-22729-0_5

28. Leung, C.-S., MacKinnon, R.K.: BLIMP: a compact tree structure for uncertain frequent pattern mining. In: Bellatreche, L., Mohania, M.K. (eds.) DaWaK 2014. LNCS, vol. 8646, pp. 115–123. Springer, Cham (2014). https://doi.org/10.1007/978-3-319-10160-6_11

29. Roy, K.K., Moon, M.H.H., Rahman, M.M., Ahmed, C.F., Leung, C.K.: Mining sequential patterns in uncertain databases using hierarchical index structure. In: Karlapalem, K., et al. (eds.) PAKDD 2021, Part II. LNCS (LNAI), vol. 12713, pp. 29–41. Springer, Cham (2021). https://doi.org/10.1007/978-3-030-75765-6_3

30. Ishita, S.Z., Ahmed, C.F., Leung, C.K.: New approaches for mining regular high utility sequential patterns. Appl. Intell. **52**, 3781–3806 (2022). https://doi.org/10.1007/s10489-021-025 36-7
31. Nguyen, H., et al.: Mining frequent weighted utility itemsets in hierarchical quantitative databases. Knowl. Based Syst. **237**, 107709:1–107709:13 (2022)
32. Nouioua, M., et al.: FHUQI-Miner: fast high utility quantitative itemset mining. Appl. Intell. **51**(10), 6785–6809 (2021). https://doi.org/10.1007/s10489-021-02204-w
33. Leung, C.K., Zhang, H., Souza, J., Lee, W.: Scalable vertical mining for big data analytics of frequent itemsets. In: Hartmann, S., Ma, H., Hameurlain, A., Pernul, G., Wagner, R.R. (eds.) DEXA 2018, Part I. LNCS, vol. 11029, pp. 3–17. Springer, Cham (2018). https://doi.org/10. 1007/978-3-319-98809-2_1
34. Zaki, M.J.: Scalable algorithms for association mining. IEEE TKDE **12**(3), 372–390 (2000)
35. Zaki, M.J., Gouda, K.: Fast vertical mining using diffsets. In: ACM KDD, pp. 326–335 (2003)
36. Agrawal, R., et al.: Mining association rules between sets of items in large databases. In: ACM SIGMOD, pp. 207–216 (1993)
37. Agrawal, R., Srikant, R.: Fast algorithms for mining association rules. In: VLDB, pp. 487–499 (1994)
38. Shenoy, P., et al.: Turbo-charging vertical mining of large databases. In: ACM SIGMOD, pp. 22–33 (2000)
39. Hsu, P.Y., et al.: Algorithms for mining association rules in bag databases. Inf. Sci. **166**(1–4), 31–47 (2004)
40. Srikant, R., Agrawal, R.: Mining quantitative association rules in large relational tables. In: ACM SIGMOD, pp. 1–12 (1996)
41. Dua, D., Graff, C.: UCI Machine Learning Repository. http://archive.ics.uci.edu/ml

Enhanced Sliding Window-Based Periodic Pattern Mining from Dynamic Streams

Evan W. Madill[1], Carson K. Leung[1]([⊠]) [ID], and Justin M. Gouge[1,2]

[1] University of Manitoba, Winnipeg, MB, Canada
Carson.Leung@UManitoba.ca
[2] York University, Toronto, ON, Canada

Abstract. Discovering frequent patterns has been an important problem for knowledge discovery. The efficient discovery of interesting patterns—such as weighted periodic patterns—from big data has been crucial to the development in new domains. Due to their high velocity of data generation and collection, these big data can form dynamic streams, which can be unbounded. Traditional approaches to this problem consist of the reconstruction of the underlying structure, while recent advances have shown new methods for dynamically updating the underlying structure for each new window. In this paper, we present an enhanced sliding window-based algorithm for mining weighted periodic patterns from dynamic streams. Evaluation results show the effectiveness of this algorithm.

Keywords: Data mining · Periodic pattern · Stream mining · Sliding window

1 Introduction

In the current era, big data [1] are everywhere. With advances in technology, high volumes of a wide variety of data are generated and collected at a high velocity for numerous real-life applications and services (e.g., healthcare informatics [2], transportation analytics [3–5], business analytics [6, 7], social network analysis [8–12]). Embedded in these big data is implicit, previously unknown and potentially useful information and knowledge. This calls for big data management [13–15], as well as big data analytics and knowledge discovery [16, 17].

As an important big data analytics and knowledge discovery task, *frequent pattern mining* [18–21] aims to discover frequently occurring sets of items (e.g., merchandise items, events) from big data. Due to the continuous and unbounded nature of dynamic data streams, their contents are usually captured in an underlying structure such as a suffix tree [22], from which frequent patterns can be mined. Groups of data (e.g., sequence of characters or a string) are usually discretized and represented by a single symbol (e.g. a character) in the tree. Traditional approaches for data stream mining with sliding windows reconstruct a suffix tree for every sliding of the windows, which can be costly. To deal with this problem, a *dynamic tree based solution to handle sliding window in time (DTSW)* [23] was proposed to dynamically update and maintain the structure of the tree for each modified window, keeping it suitable for pattern mining. Although the DTSW

R. Wrembel et al. (Eds.): DaWaK 2022, LNCS 13428, pp. 234–240, 2022.
https://doi.org/10.1007/978-3-031-12670-3_20

algorithm avoids reconstruction of suffix trees whenever the window slides, it introduces another problem. When the window slides, the deletion module of the algorithm removes the old batch from an explicit form of the tree, and insertion module inserts the new batch to an implicit form of the tree. Hence, these insertions and deletions requires frequent transformation of the tree between its implicit and explicit forms.

In this paper, we present a new algorithm to address this problem of tree transformations between its implicit and explicit forms. The algorithm eliminates the need to transform the suffix tree leaving the tree in its implicit form at all times when the window slides. Evaluation results show that our algorithm achieves a large performance increase across all window sizes tested, with no significant increase in memory.

Key contributions of this paper include design of our enhanced sliding window-based algorithm for mining periodic patterns, which are sequences that periodically occur at least a certain amount of times. With our suffixList structure, our algorithm only needs to maintain the *implicit* form of the suffix tree when capturing important information from dynamic streams (rather than converting back-and-forth between the implicit and explicit forms of the suffix tree as in the related works).

The remainder of this paper is organized as follows. The next section provides background and related works. Section 3 describes our enhanced sliding window-based algorithm for mining periodic patterns from streams. Evaluation results are shown in Sect. 4. Finally, conclusions are drawn in Sect. 5.

2 Background and Related Works

A **suffix tree** [22] is a trie containing all the suffixes of a given sequence of characters, or string. Figure 1 shows suffix trees—(a) in its implicit form and (b) in its explicit form—for a string "abcabababc". A suffix tree is in its explicit form when all suffixes can be found by traversal from the root to a leaf node. An implicit suffix tree may contain suffixes (which are implicit to an edge) that is they do not end in a leaf node, but rather end within an edge. We can force a suffix tree to be in an explicit form by inserting a unique character, usually a "$" or "#" to the end of the string.

Ukkonen's algorithm [24] is a linear-time algorithm for the construction of an implicit suffix tree. One can add a unique symbol on the end during construction to create an explicit tree. See Fig. 1.

A **sliding window** only stores data relevant to a certain time frame. Since we use the process of discretization to obtain a sequence of characters, we can form a "window" around the characters we want to look at. These characters then make up the underlying suffix tree. Consider an example with the string "abcabababc" and a window size of 3:

abc ababbc → a bca bababc → ab cab ababc

where the characters proceeding the boxed characters (i.e., characters currently in the window) may not have been received yet. In existing approaches, these suffix trees were reconstructed at each window slide.

The **Dynamic Tree Based Solution to Handle Sliding Window in Time Series Data (DTSW)** [23] handles updating of the suffix tree (rather than reconstruction) by

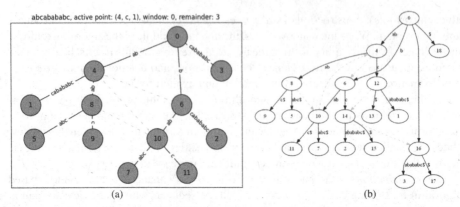

Fig. 1. A suffix tree in its (a) implicit form (as plotted by Matplotlib) and (b) explicit form (as plotted by Pydot).

the use of the insertion and deletion modules. The deletion module deletes characters (the first character) from the beginning of the window. However, before deleting any suffix from the suffix tree, the tree must be in its explicit form. To insert a new character (representing an event) onto the end of the sequence, the insertion module needs to revert back the tree from its explicit form to an implicit form for removing the unique symbol and erasing all the effects created due to it. In other words, to insert into (or delete events from) the suffix tree, the insertion and deletion modules need to convert the suffix tree back-and-forth between its explicit and implicit form.

3 Our Sliding Window-Based Weighted Periodic Pattern Mining Algorithm

The conversion between implicit and explicit form described by the DTSW is an unnecessary step. Hence, we *keep the tree in its implicit form* by proposing the idea of a *suffixList*, an ordered list of all possible suffixes from the sequence of characters. The initial tree is built with Ukkonen's algorithm and the suffixList is created during construction. We must also be able to maintain the active point in order to use Ukkonen's algorithm for insertion.

While the initial tree is being constructed using Ukkonen's algorithm, we create a **suffixList**—which is a list of all possible suffixes from the input string. We keep the list in a sorted order so that it can be easily viewed and manipulated. When deleting a character from S, we update our suffixList by removing the longest suffix. Then, we do a simple traversal of the suffix tree, and remove the nodes associated with longest suffix. In other words, we can use the suffixList to check what is supposed to be there and what is not. The process is the same for both insertion and deletion. Essentially, we use our updated suffixList, and check to make sure the necessary nodes are in the tree. Since this algorithm is meant to keep the tree in implicit form only, there is no conversion step from implicit to explicit for both insertion and deletion.

More specifically, the **deletion module** first deletes the longest suffix. Then, it traverses the tree (by traversing only the portion that starts with the character being removed). On the suffixList, it checks each suffix that starts with the character to be deleted. It goes down this tree path and updates the labels used by Ukkonen's algorithm if edge compression was used. It uses the existing structure of the tree and updates the labels as necessary, and then deletes any excess nodes that would have in the end.

A special case for deletion would be when the deletion of the first character causes two branches of the suffix tree to merge. For example, in string "abcbcbc", if it deletes 'a' from the input string, then the branch of the tree with 'a' will be removed and subtree under 'a' will be merged appropriately with 'b' branch in the tree. Otherwise, it simply traverses down the appropriate branch and updates the labels of nodes to match the suffixes as needed. Then, anything past the end will be deleted from the suffix tree.

The deletion module removes the longest suffix from the tree. Since the tree is in its implicit form, this raises the problem of searching edges that may contain smaller implicit suffixes within them as simply deletion of the longest paths leaf is not sufficient. To check how much of the edge to delete, the removal of the longest suffix from the suffixList is required. Then, a traversal checks all suffixes starting with the character being deleted. Upon a node deletion, a cleanup must be performed. Otherwise, we find the index of the longest and update the edge label to reflect the next longest matching suffix along that path.

Moreover, it also has to update the active point if necessary. When deleting the longest suffix, if the active node and edge is present when traversing to the longest suffix, then it simply deletes to the active length. Otherwise, it removes the leaf safely. If it does delete on an active point, then it decrements the remainder and finds the new active point from the remainder.

Since we maintain the active point and remainder in the deletion module, the **insertion module** simply runs Ukkonen's algorithm on the new character.

4 Evaluation

To evaluate our algorithm, we compared with reconstruction and the existing DTSW algorithm (which all implemented in Python). See Table 1 for differences among the three algorithms. In the evaluation, we used 40 window slides in order to observe the effect of how we may receive data in a stream. All tests were run on a Ryzen 7 3800X 8-Core (3.9 GHz) and 64 GB RAM (3600 MHz). We used several real-world test datasets from the UCI Machine Learning Repository[1]. As results were consistent, we reported the results—which were an average of 50 executions—for the *individual household electric power consumption dataset*, which captures 2,075,259 events discretized into 13 types.

---

[1] https://archive.ics.uci.edu/ml/index.php.

Table 1. Summary of algorithms.

Algorithm	Delete form	Insert form
Reconstruct	–	–
DTSW	Explicit	Implicit
Our algorithm	**Implicit**	Implicit

When window size grew from 10 to 1,000 and 10,000, the amount of *runtimes* required by our algorithm dropped from 42% to 22% and 15% of the runtimes required by DTSW. In contrast, for the baseline reconstruction approach (denoted as Reconstruct), runtimes grew from 4.38 ms to 793 ms and 7,950 ms when window size grew from 10 to 1,000 and 10,000, respectively. In terms of memory, as the window size grows larger, the number of characters remains the same. The *space efficiency* of the algorithms seem to, on average, converge. For instance, with 40 windows, our algorithm consumed 97% and 99% of those required by DTSW when window size = 1,000 and 10,000, respectively. See Fig. 2. To summarize, our algorithm consumes almost the same amount of memory space as the existing DTSW (baseline), but our algorithm runs much faster than DTSW.

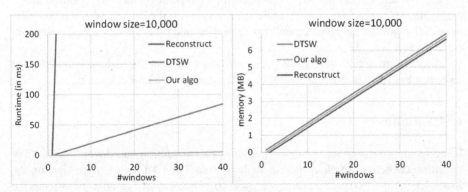

Fig. 2. (a) Runtime (in ms) and (b) memory (in MB) for sliding window performed over 40 windows with a window size of 10,000.

5   Conclusions

In this paper, we presented an enhanced sliding window-based periodic pattern stream mining algorithm. It uses suffix tree to capture important contents of the dynamic streams, from which periodic patterns are mined. It makes good use of implicit forms of the suffix tree during deletion and insertion of tree branches due to the sliding of the window capturing batches of the data streams. As such, it achieves shorter runtime and less memory space consumption when compared with the existing dynamic tree based solution to handle sliding window in time (DTSW) algorithm. Evaluation results show that our algorithm outperformed related works. As *ongoing and future work*, we explore ways (e.g., use the sliding suffix tree [25]) to further improve our algorithm.

Acknowledgement. This work is partially supported by NSERC (Canada) & U. Manitoba.

References

1. Bemarisika, P., Totohasina, A.: ERAPN, an algorithm for extraction positive and negative association rules in big data. In: Ordonez, C., Bellatreche, L. (eds.) DaWaK 2018. LNCS, vol. 11031, pp. 329–344. Springer, Cham (2018). https://doi.org/10.1007/978-3-319-98539-8_25
2. Leung, C.K., Fung, D.L.X., Hoi, C.S.H.: Health analytics on COVID-19 data with few-shot learning. In: Golfarelli, M., Wrembel, R., Kotsis, G., Tjoa, A.M., Khalil, I. (eds.) DaWaK 2021. LNCS, vol. 12925, pp. 67–80. Springer, Cham (2021). https://doi.org/10.1007/978-3-030-86534-4_6
3. Audu, A.-R.A., Cuzzocrea, A., Leung, C.K., MacLeod, K.A., Ohin, N.I., Pulgar-Vidal, N.C.: An intelligent predictive analytics system for transportation analytics on open data towards the development of a smart city. In: Barolli, L., Hussain, F.K., Ikeda, M. (eds.) CISIS 2019. AISC, vol. 993, pp. 224–236. Springer, Cham (2020). https://doi.org/10.1007/978-3-030-22354-0_21
4. Leung, C.K., Braun, P., Hoi, C.S.H., Souza, J., Cuzzocrea, A.: Urban analytics of big transportation data for supporting smart cities. In: Ordonez, C., Song, I.-Y., Anderst-Kotsis, G., Tjoa, A.M., Khalil, I. (eds.) DaWaK 2019. LNCS, vol. 11708, pp. 24–33. Springer, Cham (2019). https://doi.org/10.1007/978-3-030-27520-4_3
5. Leung, C.K., Braun, P., Pazdor, A.G.M.: Effective classification of ground transportation modes for urban data mining in smart cities. In: Ordonez, C., Bellatreche, L. (eds.) DaWaK 2018. LNCS, vol. 11031, pp. 83–97. Springer, Cham (2018). https://doi.org/10.1007/978-3-319-98539-8_7
6. Ahn, S., et al.: A fuzzy logic based machine learning tool for supporting big data business analytics in complex artificial intelligence environments. In: FUZZ-IEEE 2019, pp. 1259–1264
7. Morris, K.J., et al.: Token-based adaptive time-series prediction by ensembling linear and non-linear estimators: a machine learning approach for predictive analytics on big stock data. In: IEEE ICMLA 2018, pp. 1486–1491
8. Braun, P., Cuzzocrea, A., Jiang, F., Leung, C.K.-S., Pazdor, A.G.M.: MapReduce-based complex big data analytics over uncertain and imprecise social networks. In: Bellatreche, L., Chakravarthy, S. (eds.) DaWaK 2017. LNCS, vol. 10440, pp. 130–145. Springer, Cham (2017). https://doi.org/10.1007/978-3-319-64283-3_10
9. Jiang, F., Leung, C.K.-S.: Mining interesting "following" patterns from social networks. In: Bellatreche, L., Mohania, M.K. (eds.) DaWaK 2014. LNCS, vol. 8646, pp. 308–319. Springer, Cham (2014). https://doi.org/10.1007/978-3-319-10160-6_28
10. Leung, C.K.: Mathematical model for propagation of influence in a social network. In: Alhajj, R., Rokne, J. (eds.) Encyclopedia of Social Network Analysis and Mining, 2nd edn., pp. 1261–1269. Springer, New York, NY (2018). https://doi.org/10.1007/978-1-4939-7131-2_110201
11. Leung, C.K., et al.: Parallel social network mining for interesting "following" patterns. Concurrency Comput. Pract. Experience **28**(15), 3994–4012 (2016)
12. Leung, C.K.-S., Carmichael, C.L., Teh, E.W.: Visual analytics of social networks: mining and visualizing co-authorship networks. In: Schmorrow, D.D., Fidopiastis, C.M. (eds.) FAC 2011. LNCS (LNAI), vol. 6780, pp. 335–345. Springer, Heidelberg (2011). https://doi.org/10.1007/978-3-642-21852-1_40

13. Arora, N.R., Lee, W., Leung, C.K.-S., Kim, J., Kumar, H.: Efficient fuzzy ranking for keyword search on graphs. In: Liddle, S.W., Schewe, K.-D., Tjoa, A.M., Zhou, X. (eds.) DEXA 2012, Part I. LNCS, vol. 7446, pp. 502–510. Springer, Heidelberg (2012). https://doi.org/10.1007/978-3-642-32600-4_38

14. Eom, C.S., et al.: Effective privacy preserving data publishing by vectorization. Inf. Sci. **527**, 311–328 (2020)

15. Olawoyin, A.M., Leung, C.K., Choudhury, R.: Privacy-preserving spatio-temporal patient data publishing. In: Hartmann, S., Küng, J., Kotsis, G., Tjoa, A.M., Khalil, I. (eds.) DEXA 2020, Part II. LNCS, vol. 12392, pp. 407–416. Springer, Cham (2020). https://doi.org/10.1007/978-3-030-59051-2_28

16. Leung, C.K.-S., Jiang, F.: Big data analytics of social networks for the discovery of "following" patterns. In: Madria, S., Hara, T. (eds.) DaWaK 2015. LNCS, vol. 9263, pp. 123–135. Springer, Cham (2015). https://doi.org/10.1007/978-3-319-22729-0_10

17. Souza, J., Leung, C.K., Cuzzocrea, A.: An innovative big data predictive analytics framework over hybrid big data sources with an application for disease analytics. In: Barolli, L., Amato, F., Moscato, F., Enokido, T., Takizawa, M. (eds.) AINA 2020. AISC, vol. 1151, pp. 669–680. Springer, Cham (2020). https://doi.org/10.1007/978-3-030-44041-1_59

18. Jiang, F., Leung, C.K.-S.: Stream mining of frequent patterns from delayed batches of uncertain data. In: Bellatreche, L., Mohania, M.K. (eds.) DaWaK 2013. LNCS, vol. 8057, pp. 209–221. Springer, Heidelberg (2013). https://doi.org/10.1007/978-3-642-40131-2_18

19. Leung, C.K.-S., MacKinnon, R.K.: Balancing tree size and accuracy in fast mining of uncertain frequent patterns. In: Madria, S., Hara, T. (eds.) DaWaK 2015. LNCS, vol. 9263, pp. 57–69. Springer, Cham (2015). https://doi.org/10.1007/978-3-319-22729-0_5

20. Leung, C.-S., MacKinnon, R.K.: BLIMP: a compact tree structure for uncertain frequent pattern mining. In: Bellatreche, L., Mohania, M.K. (eds.) DaWaK 2014. LNCS, vol. 8646, pp. 115–123. Springer, Cham (2014). https://doi.org/10.1007/978-3-319-10160-6_11

21. Leung, C.K.-S., Tanbeer, S.K.: Mining popular patterns from transactional databases. In: Cuzzocrea, A., Dayal, U. (eds.) DaWaK 2012. LNCS, vol. 7448, pp. 291–302. Springer, Heidelberg (2012). https://doi.org/10.1007/978-3-642-32584-7_24

22. Rasheed, F., Alshalalfa, M., Alhajj, R.: Efficient periodicity mining in time series databases using suffix trees. IEEE TKDE **23**(1), 79–94 (2011)

23. Rizvee, R.A., et al.: Sliding window based weighted periodic pattern mining over time series data. In: ICDM 2019, pp. 118–132

24. Ukkonen, E.: On-line construction of suffix trees. Algorithmica **14**(3), 249–260 (1995)

25. Brodnik, A., Jekovec, M.: Sliding suffix tree. Algorithms **11**, 118:1–118:11 (2018)

Explainable Recommendations for Wearable Sensor Data

Niccoló Marastoni[✉], Barbara Oliboni, and Elisa Quintarelli

Department of Computer Science, University of Verona, Verona, Italy
{niccolo.marastoni,barbara.oliboni,elisa.quintarelli}@univr.it

Abstract. This paper presents an approach to explore sensor data and learn rules based on the patterns detected in the data. Our approach is a direct modification of the Apriori algorithm with a lookback mechanism that allows us to consider specific temporal windows. The inferred knowledge can be used to provide users with predictions based on historical data as well as personalized, explainable recommendations towards achieving a goal.

Keywords: Activity recommendations · Data mining · Wearable sensors data

1 Introduction

With the rapid proliferation of mobile technology, sensors and mobile devices have become ubiquitous in every aspect of our lives. They can range from wearable devices capturing the movements and heartbeats of their users, to IoT devices embedded in common appliances in homes, offices and vehicles. This guarantees that there is a constant stream of new (sensor) data that can be leveraged to extract interesting, but often implicit, information.

Wearable devices, such as Fitbit, have improved the way data can be collected from a population, providing insights on the lifestyle of their users while allowing the constant monitoring of their health-related conditions. For example, wearable devices have been successfully used to collect data on blood glucose levels in conjunction with physical activity to analyze hyperglycemic episodes in diabetic patients [5]. The large amounts of collected data can help people become more conscious of their health and aware of the possibilities to make lifestyle improvements. However, it is not always easy to obtain useful insights from these huge and semantically rich datasets. Fitbit data has been extensively used in literature to examine which factors affect the physical activity habits of college students [7], to study loneliness and social isolation [4] and even to predict blood glucose levels in diabetic patients [3].

One important aspect of sensor data is that it is intrinsically temporal in nature, since they capture events that happen in succession, labeling them with a timestamp that indicates the exact time they occurred. Indeed, it is important

R. Wrembel et al. (Eds.): DaWaK 2022, LNCS 13428, pp. 241–246, 2022.
https://doi.org/10.1007/978-3-031-12670-3_21

to know not only which events have happened, but how long they lasted, and in which order they occurred.

As a general use case we consider an application that records the daily physical activity data from Fitbit and correlates it with the sleep score, which is also obtained from Fitbit. Various analytical methods have been developed to predict the sleep quality: deep learning models have been successfully used to link daily activity rates with sleep quality [6], although some studies have reported that the correlation between physical activity and sleep quality is very weak and they might be more independent than originally thought [2].

We propose a suitable algorithm that analyzes past data, then learns which sequences of activities, along with their intensity, historically lead to our goal and tries to suggest which set of actions is best to take next in order to have a good night's sleep, as well as the actions that should be avoided. The proposed algorithm is extension of Apriori [1], which has been successfully used to infer which types of activities are associated with different levels of loneliness [4]. In our proposal, we aim at providing *explainable recommendations* [8], i.e., suitable recommendations together with intuitive and understandable explanations, in order to guarantee interpretability. The recommendation of activities to be performed during the day has to be a personalized and dynamic process tailored to a specific user. For that reason, we consider the frequent behaviours of each individual user and not those of a community of users.

The main contributions of the paper are: (i) the extension of Apriori to mine frequent sequential rules taking into account a relative notion of time (w.r.t. the current instant), and discover association rules that correlate past events with a specified future goal, and (ii) an algorithm to make both positive and negative recommendations geared towards a parametric goal.

2   Background and Motivation

Apriori [1] is a well known algorithm for finding frequent itemsets from transactional datasets; it reduces the search space by following the consideration that all non-empty subsets of a frequent itemset must be frequent. Association rules are represented as implications in the form $X \Rightarrow Y$, where X and Y are two arbitrary sets of data items such that $X \cap Y = \emptyset$. The quality of an association rule is usually measured by means of *support* and *confidence*. Support corresponds to the frequency of the set $X \cup Y$ in the dataset, while confidence corresponds to the conditional probability of finding Y, having found X and is given by $sup(X \cup Y)/sup(X)$.

In our scenario we are interested in sequential pattern mining to identify sequences of behaviours, labelled with a relative time unit, that occur frequently and are correlated with good quality sleep during the current day. More formally, a sequence is an ordered list of elements $s = \langle e_1, e_2, \ldots, e_n \rangle$, where each $e_t = \{i_1, \ldots, i_k\}$ is a collection of (unordered) items, related to the time unit t, of physical activities or sleeping activity. The length of the sequence is the number of elements in the sequence; in our use case the time unit is an interval composed of 24 hours of Fitbit tracking (i.e., the fitness activities carried out during the day and the subsequent sleeping period). For simplicity we will call the time unit *day*, thus, a sequence with cardinality n refers to the activities tracked by Fitbit in a window of n days. We call k-sequence a sequence containing k items, each of them labelled with the related time unit. The aim of sequential pattern mining is to find all subsequences of a set of sequences with support greater than, or equal to, a given threshold.

Figure 1 shows a simplified version of the Fitbit log, reporting, for each day the user's activities such as heavy physical activities (HA), light physical activities (LA), steps (ST), sleeping activity (SL) and their related intensities. The duration of each (daily) activity has been discretized into 3 possible uniform values (1: Low, 2: Medium, 3: Intense). Each day has associated in the log a transaction of items, as the activities themselves are not ordered due to the aggregated nature of Fitbit data.

Transactions	
t0	HA:2, LA:1, ST:3, SL:2
t1	HA:3, LA:2, ST:3, SL:3
t2	HA:3, LA:1, SL:2
t3	HA:3, LA:2, ST:3, SL:1
t4	LA:1, ST:3, SL:2
t5	HA:1, ST:2, SL:3
t6	HA:3, LA:3, SL:3
t7	HA:3, SL:3
t8	HA:3, LA:3, ST:1, SL:1
t9	HA:2, LA:2, ST:3, SL:2

Day-2	Day-1	Today
		HA:2, LA:1, ST:3, SL:2
	HA:2, LA:1, ST:3, SL:2	HA:3, LA:2, ST:3, SL:3
HA:2, LA:1, ST:3, SL:2	HA:3, LA:2, ST:3, SL:3	HA:3, LA:1, SL:2
HA:3, LA:2, ST:3, SL:3	HA:3, LA:1, SL:2	HA:3, LA:2, ST:3, SL:1
HA:3, LA:1, SL:2	HA:3, LA:2, ST:3, SL:1	LA:1, ST:3, SL:2
HA:3, LA:2, ST:3, SL:1	LA:1, ST:3, SL:2	HA:1, ST:2, SL:3
LA:1, ST:3, SL:2	HA:1, ST:2, SL:3	HA:3, LA:3, SL:3
HA:1, ST:2, SL:3	HA:3, LA:3, SL:3	HA:3, SL:3
HA:3, LA:3, SL:3	HA:3, SL:3	HA:3, LA:3, ST:1, SL:1
HA:3, SL:3	HA:3, LA:3, ST:1, SL:1	HA:2, LA:2, ST:3, SL:2

Fig. 1. Fitbit log of a single user   **Fig. 2.** Fitbit log related to a sequence of 3 days

Figure 2 shows the activities performed during 3 consecutive days. The column labeled "Today" is the log of all the data acquired from Fitbit up to today and is a direct transposition of the data in Fig. 1. The last element of the column is the data extracted today, while the element immediately above is the data from the previous day and so on. The column "Day-1" is the data acquired until the day before today and it starts from the second line so that every element maintains the temporal relationship with each element of the next column. Intuitively, each row is also temporally ordered just like the columns, as each element on line i of column k contains the data from the day subsequent to the element on line i of column $k - 1$.

3   Mining Frequent Patterns for Recommending Activities

Given a transactional dataset D (e.g. the Fitbit log in Fig. 1), where each transaction has a timestamp i, $D_0 \equiv D$ (the current dataset shown in Fig. 1), D_1

extends D_0 looking back 1 time unit, i.e., each transaction $t'_i \in D_1$ is the concatenation of t_{i-1} (if it exists) and t_i in D_0, and so on. In this way, we iteratively build a dataset D_j, where each transaction t is related to a time window of length $j + 1$ (as shown in Fig. 2). We can iteratively apply Apriori on the dataset D_j (looking back of j time units) and in this case our algorithm analyzes a dataset with a variable time window, or we can set the length of time window, e.g., n, and apply Apriori on the dataset D_{n-1}. After mining the frequent itemsets, we set a confidence threshold `minconf` and discover the association rules r with confidence greater than or equal to `minconf` such that the consequent is the itemset related to the target function at the time unit 0 (i.e., the current day). Indeed, we are interested in determining the user's past behaviours that influence the values of a target function, which in our running example is the user's sleeping quality.

The pseudocode of our LookBackApriori algorithm is the following:

Algorithm 1. The LookBackApriori

D_j: dataset looking back of j time units
C_k: candidate itemsets of size k
L_k: frequent itemsets of size k
$minsupp$: support threshold
for $(j = 0; L_k \neq \emptyset; j + +)$ **do**
    for $(k = 1; L_k \neq \emptyset; k + +)$ **do**
        for each transaction t in Database D_j **do**
            increment the count of all candidates in C_{k+1} that
            are contained in t
        end for
        L_{k+1} candidates in C_{k+1} with $supp > minsupp$
    end for
end for
return $\bigcup_k L_k$

The correlation between daily fitness activities and sleeping quality is represented with sequences of measurements $\mathcal{MS} = \{I_{-n}, \ldots, I_{-2}, I_{-1}, I_0\}$, where 0 is the current day, $-k$ refers to k time units before, and each I_{-k} is an itemset of measurements, which have been discretized during a preprocessing phase. We use the day as the time unit, which is comprised of the entire day's physical activities and the subsequent sleeping period. Note that a measurement sequence $\mathcal{MS}$ may be incomplete, that is, it could lack some itemsets I_{-k} (e.g., if the user was not wearing the Fitbit at time unit $-k$).

We use the measurement sequences in our Fitbit scenario to study how historical fitness and sleeping behaviours may influence the current day's sleeping quality and to recommend how to improve it or prevent it from getting worse during the current day. For this purpose we need to be able to distinguish measurements related to fitness activities (I^f_{-k}) or to sleeping activities (I^s_{-k}).

With data mining we infer association rules of the form

$$r_i : I_{-n} \wedge \cdots \wedge I_{-2} \wedge I_1 \rightarrow I_0^s \; [s_i, c_i]$$

that shows, with support s_i and confidence c_i, the correlation between the sleeping quality (i.e., our target function) for the current day 0 (see the consequent of the rule) and the activities performed looking back at most n days.

For example, a mined rule stating that after a day with medium heavy activity ($HA : 2$) and high heart rate ($HR : 3$) and the subsequent day with low light activity ($LA : 1$), the predicted sleeping quality for the current day will likely be medium ($SL : 2$), has the form

$$r : \{HA : 2, HR : 3\}_{-2} \wedge \{LA : 1\}_{-1} \rightarrow \{SL : 2\}_0 \; [s_r, c_r]$$

During the recommendation step of our methodology we build two sets of rules $\mathcal{R}^+(\bar{r})$ and $\mathcal{R}^-(\bar{r})$, that contain the recommendations on the fitness activities that may lead to better sleeping quality and to worse sleeping quality, respectively. On the current day 0 we can access the user's Fitbit log and obtain the data collected during the past n days, i.e., $L = \langle I'_{-n}, \ldots, I'_{-2}, I'_{-1} \rangle$.

For example, during the previous 3 days the user log may be:

$$L = \langle \{LA : 1, ST : 3, SL : 2\}_{-3}, \{HA : 3, ST : 2, SL : 3\}_{-2},$$
$$\{HA : 3, LA : 3, SL : 3\}_{-1} \rangle$$

The rule $\bar{r}$ is the best rule for answering the query *"Will I sleep well tonight?"* and is of the form $\bar{r} : I_{-n} \wedge \cdots \wedge I_{-2} \wedge I_{-1} \rightarrow I_0^s$, with $supp > s_{threshold}$ and maximum confidence c w.r.t. other rules of the same form, and such that $\bar{r} \subseteq L$. Note that for the current time unit 0 the antecedent does not contain any itemset related to the fitness activity (i.e., I_0^f) because we assume the recommendations are useful at the beginning of the time unit 0. If this is not the case, we could mine rules having also an itemset I_0^f in the antecedent.

We say that a rule r is contained in the user log L (i.e., $r \subseteq L$) iff $\forall i \in \{1, \ldots, n\} \; I_{-i} \subseteq I'_{-i}$, that is, every itemset in the rule is contained in what the user has done in the past n days and stored in his/her log. Note that independently from the log, a rule may be incomplete and contain itemsets only for some days of the considered window of length n.

For example, the rule $\bar{r}_1 : \{HA : 3\}_{-2} \wedge \{HA : 3\}_{-1} \rightarrow \{SL : 2\}_0$ is contained in the log L, whereas the rule $\bar{r}_2 : \{HA : 3\}_{-3} \wedge \{HA : 3\}_{-2} \wedge \{HA : 3\}_{-1} \rightarrow \{SL : 2\}_0$ is not contained in L because it is not true that the user performed intense heavy activity 3 days ago. In this way, when the users receive the rule $\bar{r}_1$, in the antecedent they can find an explanation of the reason why their predicted sleep quality is medium.

The two sets of recommendations are:

$$\mathcal{R}^+(\bar{r}) = \{r \mid r : I_{-n} \wedge \cdots \wedge I_{-2} \wedge I_{-1} \wedge \mathbf{I_0^f} \rightarrow \widetilde{I}_0^s \text{ with } \widetilde{I}_0^s > I_0^s\}$$

$$\mathcal{R}^-(\bar{r}) = \{r \mid r : I_{-n} \wedge \cdots \wedge I_{-2} \wedge I_{-1} \wedge \mathbf{I_0^f} \rightarrow \widetilde{I}_0^s \text{ with } \widetilde{I}_0^s < I_0^s\}$$

The set $\mathcal{R}^+$ is composed of the rules with the same past activities of $\overline{r}$, but with a suggestion of fitness activities for the current day (i.e., I_0^f) and with better sleeping quality in the consequent (i.e., $\widetilde{I}_0^s$ such that $\widetilde{I}_0^s > I_0^s$). On the contrary, the rules in $\mathcal{R}^-$ are those with the same past activities of $\overline{r}$, but with a suggestion of fitness activities for the current day (i.e., I_0^f) that may lead to worse sleeping quality in the consequent. The order relation $>$ depends on the function we want to optimize.

An example of a negative recommendation in $\mathcal{R}^-$ is the suggestion to avoid heavy activities in the current day, since after three consecutive days of intense heavy activities, the sleep quality tends to decrease.

4   Conclusions

In this paper we have introduced LookBackApriori, an algorithm to mine frequent sequences in data and generate rules geared towards a specific goal. The rules can be used to suggest actions to take and to avoid in order to reach the goal, thus, they contain an explanation as well. As future work, we plan to infer not just the frequent activity patterns, but also the average duration of each activity correlated with external information (e.g., weather conditions).

References

1. Agrawal, R., Srikant, R.: Fast algorithms for mining association rules in large databases. In: Bocca, J.B., Jarke, M., Zaniolo, C. (eds.) Proceedings of VLDB 1994, pp. 487–499. Morgan Kaufmann (1994)
2. Angelides, M.C., Wilson, L.A.C., Echeverría, P.L.B.: Wearable data analysis, visualisation and recommendations on the go using android middleware. Multimed. Tools Appl. **77**(20), 26397–26448 (2018). https://doi.org/10.1007/s11042-018-5867-y
3. Bosoni, P., Meccariello, M., Calcaterra, V., Larizza, C., Sacchi, L., Bellazzi, R.: Deep learning applied to blood glucose prediction from flash glucose monitoring and fitbit data. In: Michalowski, M., Moskovitch, R. (eds.) AIME 2020. LNCS (LNAI), vol. 12299, pp. 59–63. Springer, Cham (2020). https://doi.org/10.1007/978-3-030-59137-3_6
4. Doryab, A., et al.: Identifying behavioral phenotypes of loneliness and social isolation with passive sensing: statistical analysis, data mining and machine learning of smartphone and fitbit data. JMIR Mhealth Uhealth **7**(7), e13209 (2019)
5. Salvi, E., et al.: Patient-generated health data integration and advanced analytics for diabetes management: the AID-GM platform. Sensors **20**(1), 128 (2020)
6. Sathyanarayana, A., et al.: Sleep quality prediction from wearable data using deep learning. JMIR Mhealth Uhealth **4**(4), e125 (2016)
7. Wang, C., Lizardo, O., Hachen, D.S.: Using fitbit data to examine factors that affect daily activity levels of college students. PLOS One **16**(1), e0244747 (2021)
8. Zhang, Y., Lai, G., Zhang, M., Zhang, Y., Liu, Y., Ma, S.: Explicit factor models for explainable recommendation based on phrase-level sentiment analysis. In: ACM SIGIR Conference, pp. 83–92. ACM (2014)

Machine Learning

SLA-Aware Cloud Query Processing with Reinforcement Learning-Based Multi-objective Re-optimization

Chenxiao Wang[1]([✉]), Le Gruenwald[1], and Laurent d'Orazio[2]

[1] School of Computer Science, University of Oklahoma, Norman, OK, USA
{chenxiao,ggruenwald}@ou.edu
[2] CNRS IRISA, Rennes 1 University, Lannion, France
laurent.dorazio@univ-rennes1.fr

Abstract. Query processing on cloud database systems is a challenging problem due to the dynamic cloud environment. In cloud database systems, besides query execution time, users also consider the monetary cost to be paid to the cloud provider for executing queries. Moreover, a Service Level Agreement (SLA) is signed between users and cloud providers before any service is provided. Thus, from the profit-oriented perspective for the cloud providers, query re-optimization is multi-objective optimization that minimizes not only query execution time and monetary cost but also SLA violations. In this paper, we introduce ReOptRL and SLAReOptRL, two novel query re-optimization algorithms based on deep reinforcement learning. Experiments show that both algorithms improve query execution time and query execution monetary cost by 50% over existing algorithms, and SLAReOptRL has the lowest SLA violation rate among all the algorithms.

Keywords: Query optimization · Cloud databases · Reinforcement learning · Query re-optimization

1 Introduction

In a traditional database management system (DBMS), finding the query execution plan (QEP) with the best query execution time among those QEPs generated by a query optimizer is the key to the performance of a query. However, in a cloud database system, minimizing query response time is not the only goal of query optimization. As hardware usage is charged on-demand and scalability is available to users, query execution monetary cost also needs to be considered as one of the objectives for optimizing QEPs. Meanwhile, the cloud providers need to minimize SLA violation rate in addition to fulfilling the users' requirements of query execution time and monetary cost for query execution. Traditionally, the query optimizer evaluates the time and monetary costs of different QEPs to derive the best QEP for a query before execution. These time and monetary costs are estimated based on the data statistics available to the query optimizer at the moment when the query optimization is performed. These statistics are often approximate, which may result in inaccurate estimates for the time and monetary costs

R. Wrembel et al. (Eds.): DaWaK 2022, LNCS 13428, pp. 249–255, 2022.
https://doi.org/10.1007/978-3-031-12670-3_22

needed to execute the query. Thus, the QEP generated before query execution may not be the best one.

To deal with this issue, there exist methods proposed to re-optimize queries during their execution [1–3]. Ortiz and et al. [1] apply deep reinforcement learning (RL) to learn a representation of queries, which can then be used in downstream query optimization tasks. Marcus and et al. present work of a deep RL-based join optimizer, ReJOIN [2], which orders a preliminary view of the potential for deep RL in this context. In these techniques, QEPs are re-optimized multiple times by a deep RL model. Kipf [3] uses Deep Neural Network (DNN) to learn cardinality estimates. Wu and et al. [6] have proposed Sample, a query re-optimization algorithm that updates data statistics estimated from a sample of tuples collected during runtime. However, none of them addresses monetary costs and SLA requirements for cloud databases at the same time. In this paper, we present two algorithms, ReOptRL and SLAReOptRL, that use reinforcement learning to perform multi-objective query re-optimization for query processing in an end-to-end cloud database system. The algorithms employ a reward function designed specifically for query re-optimization considering query execution time, money cost and SLA requirements.

2   The Reinforcement Learning-Based Multi-objective Query Re-optimization Algorithm (ReOptRL)

We choose RL instead of supervised learning methods because RL does not require training data, which is a labeled dataset of past actions, to be available in advance to train the learning model before the model can be used to predict future actions. There are various kinds of RL algorithms that have been proposed. Q-Learning is one of the popular value-based RL algorithms and using the Bellman equation [4].

$$Q(S_t, a_t) \leftarrow Q(S_t, a_t) + \alpha(R_t + \Upsilon Q(S_{t+1}, a_{t+1}) - Q(S_t, a_t)) \tag{1}$$

In Q-Learning, a table (called Q-table) is used to store all the potential state-action pairs (S_n, a_n) and an evaluated Q-value associated with this pair. In Eq. (1), $Q(S_t, a_t)$ is an evaluated value (called Q-value) for executing Action a_t at State S_t. This value is used to select the best Action to perform under the current state. In our scenario, there are many available containers on which a single query operator can be executed. Thus, many state-action pairs are in the Q-table potentially. Iterating a large Q-table incurs extra time overhead which delays the query execution. To solve this issue, we apply Deep Q Network (DQN) [4] as our reinforcement learning for query re-optimization. DQN works similarly to Q-Learning. The major difference is that, given a state, instead of using the Q-table, it uses a neural network to estimate the Q-values for all the potential actions. The input of the neural network is the current state. For the current QEP to represent the current state, we use a one-hot vector adapted from the recent work [2] to represent a QEP. The ReOptRL algorithm is given in Fig. 1. First, a query is submitted to a query optimizer which generates the QEP (logical plan) for the query (Line 4). Then the QEP is converted into a one-hot vector representation (Line 7). This vector is sent to the RL model, which is a neural network. The RL model will evaluate the Q-values for all the potential actions to execute the next available query operator (Line 8). Each of these actions consists of two parts, a physical operator and a container (machine) to execute the physical operator. Then the action with the best Q-value will be selected and performed by the DBMS (Line 9). After that, the executed query operator is discarded from the QEP (Line 10). The reward is updated with the time and monetary cost needed to execute the operator and then the expected Q-value is updated by the Bellman Eq. (2) with the updated reward (Lines 11–13). The weights of the neural network are updated accordingly by the back-propagation method (Line 14). This process repeats for each operator in the QEP and terminates when all the operators in the QEP are executed. The query results are then sent to the user (Line 17).

Algorithm: Reinforcement Learning Based Multi-Objective Query Re-Optimization (ReOptRL)

INPUT: SQL query, Weight Profile wp, Reward Function R (), Learning rate α, Discount rate Υ
OUTPUT: The query result set of the input query
1.   t=0
2.   Result = Ø
3.   Q$_t$= 0
4.   QEP = QueryOptimizer(query)
5.   **while** QEP≠ Ø
6.   Op=next available operator in QEP
7.   State S$_t$= convert QEP to a state vector
8.   Action$_t$=RunLearningModel (S$_t$, wp)
9.   Result=Result ∪ execute (Op, Action$_t$)
10.   QEP=QEP-Op
11.   Update R$_N$=R (wp, Action$_t$.time, Action$_t$.money))
12.   Obtain Q-value of next state Q$_{t+1}$ from the neural network
13.   Update Q-value of current state Q$_t$ = Bellman (Q$_t$, Q$_{t+1}$, R$_t$, α, Υ)
14.   Update Weights in the neural network
15.   t=t+1
16. **end while**
17. **return** Result

Fig. 1. The ReOptRL algorithm

In ReOptRL, after an action is performed, the reward function is used to evaluate the action. This gives feedback on how the selected action performs to the learning model. The performed action with a high reward will be more likely to be selected again under the same state. The reward function plays a key role in the entire algorithm. According to the Bellman equation, if the reward of performing a previous action A$_{t-1}$ is high on state S$_{t-1}$, the Q-value will also be high. This means that, given the same state, the action with the good previous performance will have a higher chance to be selected. In our algorithm, we would like the actions with low query execution time and monetary cost to be the ones that will be more likely to be chosen. To reflect this feature, we define the reward function as follows:

$$Reward\ R = \frac{1}{1 + (W_t * T_{op}^q) + (W_m * M_{op}^q)} \tag{2}$$

where W_t and W_m are the time and monetary weights provided by the user, and T_{op}^q and M_{op}^q are the time and monetary costs for executing the current operator op in query q.

According to this reward function, the query is executed based on the user's preference.

which is either the user wanting to spend more money for a better query execution time or vice versa. We call these two preferences Weights. These two weights defined by the user are called Weight Profile (wp), which is a two-dimensional vector, and each dimension is a number between 0.0 to 1.0. Notice that the user only needs to specify one dimension of the weight profile, the other dimension is computed as 1-Weight automatically. The detail can be found in our previous work [5].

3   The SLA-Aware Reinforcement Learning-Based Multi-objective Query Re-optimization Algorithm (SLAReOptRL)

An SLA is a contract between cloud service providers and consumers, mandating specific numerical target values which the service needs to achieve. Considering an SLA in query processing is important for cloud databases. If an SLA violation happens, the cloud service providers need to pay a penalty to their users in a form such as money or CPU credits. From a profit-oriented perspective, cloud service providers would want to keep the number of SLA violations as low as possible. Different cloud service providers implement different SLAs with their users. Using time and monetary costs to execute a query as the SLA requirements has been studied in [1]. We find them practical and more specific to users and thus adopt the same SLA requirements in our work.

In particular, the reward function shown in Eq. (4) is extended from Eq. (2) to make it possible to select the best action according to the SLA requirements:

$$Reward\ R = \frac{1}{1 + \left(W_t * \left(T_{op}^q + P_t\right)\right) + \left(W_m * \left(M_{op}^q + P_m\right)\right)} \tag{3}$$

where T_{op}^q and M_{op}^q are the time and monetary costs for executing the current operator op in query q

$$P_t = \alpha_{op} * delay\_time, \quad P_m = \alpha_{op} * exceeded\_money$$

where α_{op} is the operator impact rate of the operator type op

$$delay\_time = \begin{cases} 0 \\ T_{op}^q - SLA.T_{op}^q \ if\ T_{op}^q > SLA.T_{op}^q \end{cases} \tag{4}$$

$$exceeded\_money = \begin{cases} 0 \\ M_{op}^q - SLA.M_{op}^q \ if\ M_{op}^q > SLA.M_{op}^q \end{cases} \tag{5}$$

In this reward function (Eq. (4)), P_t and P_m reflect the extra costs for executing a query operator if the SLA is violated. If the SLA is not violated for executing every operator, then this equation is the same as the reward function used in ReOptRL (Eq. (2)). In Eqs. (4) and (5), delay_time is the amount of difference between the actual time to execute a query operator and the maximum time allowed to execute this query operator as specified in the SLA. The same idea applies to *exceeded_money* for monetary costs. We use these two values to quantify the SLA violation on query execution time and monetary cost. In Eq. (4), these two values are used to compute P_t and P_m. It shows that the larger the number of SLA violations, the smaller the reward becomes. We build the reward function this way so that the reward is related to SLA violations. Also, we use the query operator impact rate α_{op} to scale up the impact of SLA violations on different types of operators.

4 Performance Evaluation

There are two sets of machines used in our experiments. A single local machine used to train the machine learning model and to perform the query optimization. This local machine has an Intel i5 2500K Dual-Core processor running at 3 GHz with 16 GB DRAM. The second set consists of 10 dedicated Virtual Private Servers (VPSs) that are used for the deployment of the query execution engine. The query optimizer and the query engine used in the experiments are modified from the open-source database management system, PostgreSQL 8.4. The data are distributed among these VPSs. The queries and database tables are generated using the TPC-H database benchmark. The database tables are populated with 1,000 GB data using the default data generator. We run 50,000 queries and these queries are generated by the query templates randomly selected from the 22 query templates from the benchmark.

We compare the performance results obtained when the following query re-optimization algorithms are incorporated into query processing: 1) our two proposed algorithms, **ReOptRL** and **SLAReOptRL**; 2) the algorithm where a query re-optimization is conducted automatically after the execution of each operator in the query is completed (denoted as **ReOpt**), which we developed based on the work presented in [5]; 3) the algorithm where a query re-optimization is conducted by a supervised machine

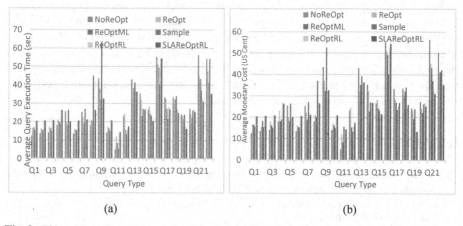

(a)   (b)

Fig. 2. Time (a) and monetary cost (b) performance for executing queries using different algorithms

learning model decision (denoted as **ReOptML**). 4) the algorithm proposed in [6] where query optimization uses sampling-based query estimation (denoted as **Sample**), and 5) the algorithm that uses no re-optimization (denoted as **NoReOpt**).

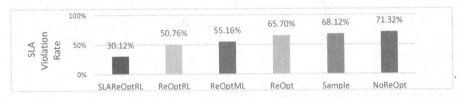

Fig. 3. Average SLA violation rates when executing queries using different algorithms

From Fig. 2 (a) and (b), we can see that, for both the query execution time and monetary costs, on average SLAReOptRL performs the best and ReOptRL performs the second best among all the algorithms. Specifically, comparing with the baseline NoReOpt where no re-optimization is conducted, the query execution time improvement using SLAReOptRL is 45%, ReOptRL 39%, ReOptML 27%, ReOpt 13%, and Sample 10%, while the monetary cost improvement using SLAReOptRL is 62%, ReOptRL 52%, ReOptML 27%, ReOpt 17%, and Sample 5%. Especially, the monetary cost has a significant improvement (SLAReOptRL and ReOptRl are 62% and 52% better than NoReOpt, repsectively). Moreover, from Fig. 3, we can also find that by using SLAReOptRL, the SLA violation rate is the lowest one among the SLA violation rates caused by all the algorithms. This shows the positive effect of considering SLA requirements in re-optimization.

5 Conclusion

This paper presents two query re-optimization algorithms called ReOptRL and SLAReOptRL. Both use a reinforcement learning-based model to decide the physical query operator and machines to execute an operator from a query execution plan (QEP) for a query in a cloud database system. The experiments conducted using the TPC-H database benchmark show that both SLAReOptRL and ReOptRL improve query response time (from 12% to 45%) and monetary cost (from 17% to 62%) over the existing algorithms In addition, we also find that when there are SLA requirements, SLAReOptRL performs 20% better than ReOptRL on SLA violation rate.

References

1. Ortiz, J., Almeida, V.T., Balazinska, M.: Changing the face of database cloud services with personalized service level agreements. In: CIDR 2015 (2015)
2. Marcus, R., Papaemmanouil, O.: Deep reinforcement learning for join order enumeration. In: aiDM 2018, pp. 1–4 (2018)
3. Kandi, M.M., Yin, S., Hameurlain, A.: An integer linear-programming based resource allocation method for SQL-like queries in the cloud. In: SAC 2018, pp. 161–166 (2018)

4. Wiering, M., Otterlo, M.V.: Reinforcement Learning: State-of-the-Art. Springer Publish-ing Company, Incorporated, Berlin, Heidelberg (2014). https://doi.org/10.1007/978-3-642-276 45-3

5. Wang, C., Arrani, Z., Gruenwald, L., Laurent, D.: Adaptive time- monetary cost aware query optimization on cloud DataBase. In: Big Data 2018, pp. 3374–3382 (2018)

6. Wu, W., Naughton, J.F., Singh, H.: Sampling-based query re-optimization. In: SIGMOD 2016, pp. 1721–1736 (2016)

Mahalanobis Distance Based K-Means Clustering

Paul O. Brown[1], Meng Ching Chiang[1], Shiqing Guo[1], Yingzi Jin[1],
Carson K. Leung[1]($\boxtimes$) (iD), Evan L. Murray[1], Adam G. M. Pazdor[1],
and Alfredo Cuzzocrea[2]

[1] University of Manitoba, Winnipeg, MB, Canada
Carson.Leung@UManitoba.ca
[2] University of Calabria, Rende, Italy

Abstract. In the current era, big data are everywhere. With advances in technology, high volumes of a wide variety of data are generated and collected in numerous real-life applications and services. Embedded in these big data is implicit, previously unknown and potentially useful information and knowledge. This calls for data science. Among various data science tasks, clustering is an important one. Although there have been techniques to improve the accuracy of k-means clustering algorithms, many of them are applied independently. In this paper, we present a k-means clustering algorithm with Mahalanobis distance. This is a non-trivial integration of partitioning based clustering, correlation based clustering, and Mahalanobis distance. Evaluation results show our algorithm is more accurate than the related works to cluster similar data.

Keywords: Data mining · Clustering · Machine learning · Unsupervised learning · k-means · Mahalanobis distance

1 Introduction

In the current era, big data [1] are everywhere. With advances in technology, high volumes of a wide variety of data are generated and collected in numerous real-life applications and services (e.g., healthcare informatics [2], transportation analytics [3–5], business analytics [6, 7], social network analysis [8–12]). Embedded in these big data is implicit, previously unknown and potentially useful information and knowledge. This calls for big data management [13–15], as well as big data analytics and knowledge discovery [16, 17].

In addition to frequent pattern mining [18–21], another important big data analytics and knowledge discovery task is *clustering* [22–25]. It aims to group similar objects together and distinguish them from dissimilar objects. It has been applied in numerous applications [26–28] such as clustering similar sounding names or text [29, 30]. *K-means* [31, 32] is a popular technique within the family of partition clustering, partially due to its simplicity and ease of use. It partitions the dataset X into k clusters (i.e., groups), and each cluster contains at least one data point. *Key contributions* of this work include our

© The Author(s), under exclusive license to Springer Nature Switzerland AG 2022
R. Wrembel et al. (Eds.): DaWaK 2022, LNCS 13428, pp. 256–262, 2022.
https://doi.org/10.1007/978-3-031-12670-3_23

design of our clustering algorithm, which incorporates Mahalanobis distance [33, 34] into k-means clustering.

The remainder of this paper is organized as follows. The next section provides background and related works. Section 3 describes our Mahalanobis distance based k-means elliptical clustering algorithms. Evaluation results are shown in Sect. 4. Finally, conclusions are drawn in Sect. 5.

2 Background and Related Work

A widely used clustering technique is the **k-means** clustering algorithm [35, 36]. It aims to partition a dataset X into k clusters, each cluster being represented by the mean value of its data points. Its key ideas can be described as follows. First, each cluster C_i (where $1 \leq i \leq k$) is initialized to be a set of points in X that are closer to the center c_i of cluster C_i than to that of other clusters. Then, for each cluster C_i, compute its new center of all the points within C_i. Afterwards, repeats the aforementioned process until c_i and C_i do not change for any cluster.

While k-means is simple and easy to use, its performance depends on the initial centroids and the distance function [37]. Initial centroids are often randomly chosen. The (dis)similarity distance is often measured by Euclidean distance (aka L^2-norm), which can be defined as:

$$dist(X, Y) = \sqrt{\sum_{j=1}^{n} \left(x_j - y_j \right)^2} \tag{1}$$

where $X = (x_1, \ldots, x_n)$ and $Y = (y_1, \ldots, y_n)$ in Euclidean space $\mathbb{R}^n$. For clustering with Euclidean distance, it aims to minimize intra-class dissimilarity:

$$\underset{C}{\text{argmin}} \sum_{i=1}^{k} \sum_{x \in C_i} \|x - \mu_i\|^2 = \underset{C}{\text{argmin}} \sum_{i=1}^{k} \left(|C_i| \text{Var}(C_i) \right) \tag{2}$$

where μ_i is the center (i.e., centroid) of C_i. Euclidean distance works well for spherical clusters where every point is within a radius, but may not do so for elliptical clusters. As some points would be significantly farther away than others, they may drift off and be categorized into the wrong cluster. Besides Euclidean distance, other popular distance functions [38] include:

- Manhattan distance (aka taxicab or L^1-norm): $dist(X, Y) = \sum_{j=1}^{n} |x_j - y_j|$
- Chebyshev distance (aka chessboard or L^∞-norm): $dist(X, Y) = \max_{j} |x_j - y_j|$

- Minkowski distance (aka L^p-norm): $dist(X, Y) = \sqrt[p]{\sum_{j=1}^{n} |x_j - y_j|^p}$. Hence, the three aforementioned distances can be considered as special cases of Minkowski distance (e.g., $p = 1$ to become Manhattan distance, $p = 2$ to become Euclidean distance, and $\lim_{p \to \infty} dist(X, Y)$ to become Chebyshev distance).

As another alternative, **Mahalanobis distance** [33, 34] takes the correlation of a dataset into account by using an inverse of a variance-covariance matrix of the dataset, which can be used to measure the number of standard deviations from one point to another. Thus, it is unitless and scale-invariant. In general, Mahalanobis distance can be defined as:

$$dist(X, Y) = \sqrt{(X - Y)^T S^{-1}(X - Y)} \tag{3}$$

where $X, Y \in \mathbb{R}^n$ and S is a positive-definite variance matrix (aka variance-covariance matrix), in which:

- diagonal entries are variance of X, i.e., $S_{XX} = \text{Var}(X)$, and
- non-diagonal entries are covariance of X and Y, i.e., $S_{XY} = \text{cov}(X, Y)$.

3 Our Clustering Algorithm

Incorporating Mahalanobis distance into the k-means algorithm is not as simple as calling Mahalanobis distance instead of Euclidean distance. One needs to calculate the inverse of the variance-covariant matrix, which cannot be calculated without clusters. This leads to our first modification of the standard k-means, which is to initialize its clusters. Two logical choices to initialize the first cluster are (a) randomly assign points to clusters, or (b) use the first iteration of Euclidean distance based k-means algorithm to select the initial clusters with given centroids. As it is vital to allow smarter centroid selection, our algorithm selects the latter of the two choices. Euclidean distance is defined as in Eq. (1). Our algorithm then calculates Euclidean distance of each point to each of the initialized centroids, and assigns the point to its closest centroid.

After initializing the clusters, our algorithm then calculates the inverse variance-covariance S^{-1} for each cluster. Note that variance-covariance matrix $S_{n \times n}$ is symmetric and positive-definite, and can be computed by:

$$S(X, Y) = \begin{cases} \text{Var}(X) = \text{cov}(X, X) = \text{E}[(X - \text{E}(X))(X - \text{E}(X))^T] & \text{if } X = Y \\ \text{cov}(X, Y) = \text{E}[(X - \text{E}(X))(Y - \text{E}(Y))^T] & \text{otherwise} \end{cases} \tag{4}$$

where $\text{E}(X)$ is the expected value (i.e., mean) of X. Next, our algorithm measures Mahalanobis distance of each point with a particular centroid point and its respective cluster inverse variance-covariance. Here, it aims to minimize:

$$\underset{C}{\text{argmin}} \sum_{i=1}^{k} \sum_{x \in C_i} \sqrt{(x - \mu_i)^T S^{-1}(x - \mu_i)} \tag{5}$$

where μ_i is the center (i.e., centroid) of C_i. Afterwards, it computes the new centroids by calculating the mean of each cluster. Lastly, it checks if the previous centroids are the same as the new centroids. If so, then it reaches convergence, and thus the clustering is completed. Otherwise, it sets the previous centroids to the current ones, and repeats the loop again. Whenever the loop starts again, the inverse variance-covariance is recalculated. Figure 1 shows a pseudo code of the resulting algorithm.

Algorithm Mahalanobis *k*-means

Input: Dataset P of points, initial centroids C, number of clusters k

Output: Labels indicating which points are in which cluster, the final centroids

```
1:  cluster = initial_cluster(C)
2:  while previous C is not the same do
3:      inverseVarCov = initial_inverseVarCov(cluster); cluster = []; labels = []
4:      for p in P do
5:          dists = []
6:          for c in C do dists.append(MahalanobisDist(p, c, inverseVarianceCovariance))
7:          labels.append(min(dists)); cluster[min(dists)].append(p)
8:      new_centroids = []
9:      for c in clusters do new_centroids.append(mean of c)
10:     check centroid and new_centroids
11: return label
```

Fig. 1. Pseudo code of our Mahalanobis *k*-means algorithm.

4 Evaluation

To evaluate our Mahalanobis distance based *k*-means algorithm, we compared our algorithm with the existing Euclidean distance based *k*-means algorithm on several datasets. Figure 2 shows visualization of two comparisons. For *spherical clustering*, both algorithms successfully grouped all data points and formed the same three clusters as shown in Fig. 2(a). However, for *elliptical clustering*, as shown in Fig. 2(b), the existing Euclidean distance based algorithm failed to cluster the elliptical dataset correctly. In contrast, our Mahalanobis distance based *k*-means algorithm clustered the data points correctly as shown in Fig. 2(c).

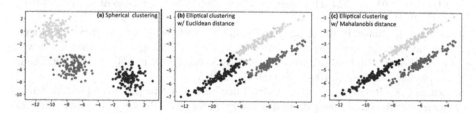

Fig. 2. Results for (a) spherical clustering from both existing Euclidean distance based and our Mahalanobis distance based *k*-means algorithm, (b) elliptical clustering from Euclidean distance based *k*-means algorithm, and (c) elliptical clustering from our Mahalanobis distance based *k*-means algorithm.

In terms of quantifiable results, our Mahalanobis distance based *k*-means algorithm took around ~7 s to run, which was faster than its Euclidean distance based counterpart. As for accuracy, our Mahalanobis distance based *k*-means algorithm achieved a Rand index [39] and V-measure [40]—which were both bounded within 0.0 to 1.0— that were 0.2 higher than its Euclidean distance based counterpart. Note that the Rand index measures the similarity between the ground-truth class assignment and our clustering algorithm assignments of the same samples. V-measure is a harmonic mean for measuring homogeneity (i.e., each cluster contains only members of a single class) and completeness (i.e., all members of a given class are assigned to the same cluster).

5 Conclusions

In this paper, we presented a non-trivial integration of k-means clustering and Mahalanobis distance. Evaluation results show our Mahalanobis distance based k-means clustering algorithm is more accurate than the related works. As *ongoing and future work*, we explore ways to further enhance our algorithm and examine impacts of Mahalanobis distance on other clustering techniques.

Acknowledgement. This work is partially supported by NSERC (Canada) & U. Manitoba.

References

1. Bemarisika, P., Totohasina, A.: ERAPN, an algorithm for extraction positive and negative association rules in big data. In: Ordonez, C., Bellatreche, L. (eds.) DaWaK 2018. LNCS, vol. 11031, pp. 329–344. Springer, Cham (2018). https://doi.org/10.1007/978-3-319-98539-8_25
2. Leung, C.K., Fung, D.L.X., Hoi, C.S.H.: Health analytics on COVID-19 data with few-shot learning. In: Golfarelli, M., Wrembel, R., Kotsis, G., Tjoa, A.M., Khalil, I. (eds.) DaWaK 2021. LNCS, vol. 12925, pp. 67–80. Springer, Cham (2021). https://doi.org/10.1007/978-3-030-86534-4_6
3. Audu, A.-R.A., Cuzzocrea, A., Leung, C.K., MacLeod, K.A., Ohin, N.I., Pulgar-Vidal, N.C.: An intelligent predictive analytics system for transportation analytics on open data towards the development of a smart city. In: Barolli, L., Hussain, F.K., Ikeda, M. (eds.) CISIS 2019. AISC, vol. 993, pp. 224–236. Springer, Cham (2020). https://doi.org/10.1007/978-3-030-22354-0_21
4. Leung, C.K., Braun, P., Hoi, C.S.H., Souza, J., Cuzzocrea, A.: Urban analytics of big transportation data for supporting smart cities. In: Ordonez, C., Song, I.-Y., Anderst-Kotsis, G., Tjoa, A.M., Khalil, I. (eds.) DaWaK 2019. LNCS, vol. 11708, pp. 24–33. Springer, Cham (2019). https://doi.org/10.1007/978-3-030-27520-4_3
5. Leung, C.K., Braun, P., Pazdor, A.G.M.: Effective classification of ground transportation modes for urban data mining in smart cities. In: Ordonez, C., Bellatreche, L. (eds.) DaWaK 2018. LNCS, vol. 11031, pp. 83–97. Springer, Cham (2018). https://doi.org/10.1007/978-3-319-98539-8_7
6. Ahn, S. et al.: A fuzzy logic based machine learning tool for supporting big data business analytics in complex artificial intelligence environments. In: FUZZ-IEEE 2019, pp. 1259–1264 (2019)
7. Morris, K.J., et al.: Token-based adaptive time-series prediction by ensembling linear and non-linear estimators: a machine learning approach for predictive analytics on big stock data. In: IEEE ICMLA 2018, pp. 1486–1491 (2018)
8. Braun, P., Cuzzocrea, A., Jiang, F., Leung, C.K.-S., Pazdor, A.G.M.: MapReduce-based complex big data analytics over uncertain and imprecise social networks. In: Bellatreche, L., Chakravarthy, S. (eds.) DaWaK 2017. LNCS, vol. 10440, pp. 130–145. Springer, Cham (2017). https://doi.org/10.1007/978-3-319-64283-3_10

9. Jiang, F., Leung, C.K.: Mining interesting "following" patterns from social networks. In: Bellatreche, L., Mohania, M.K. (eds.) DaWaK 2014. LNCS, vol. 8646, pp 308–319. Springer, Cham (2014). https://doi.org/10.1007/978-3-319-10160-6_28

10. Leung, C.K.: Mathematical model for propagation of influence in a social network. In: Alhajj, R., Rokne, J. (eds.) Encyclopedia of Social Network Analysis and Mining, 2nd edn., pp. 1261–1269. Springer, New York, NY (2018). https://doi.org/10.1007/978-1-4939-7131-2_110201

11. Leung, C.K., et al.: Parallel social network mining for interesting 'following' patterns. Concurrency Computation Pract. Experience 28(15), 3994–4012 (2016)

12. Leung, C.K.-S., Carmichael, C.L., Teh, E.W.: Visual analytics of social networks: mining and visualizing co-authorship networks. In: Schmorrow, D.D., Fidopiastis, C.M. (eds.) FAC 2011. LNCS (LNAI), vol. 6780, pp. 335–345. Springer, Heidelberg (2011). https://doi.org/10.1007/978-3-642-21852-1_40

13. Arora, N.R., Lee, W., Leung, C.K.-S., Kim, J., Kumar, H.: Efficient fuzzy ranking for keyword search on graphs. In: Liddle, S.W., Schewe, K.-D., Tjoa, A.M., Zhou, X. (eds.) DEXA 2012, Part I. LNCS, vol. 7446, pp. 502–510. Springer, Heidelberg (2012). https://doi.org/10.1007/978-3-642-32600-4_38

14. Eom, C.S., et al.: Effective privacy preserving data publishing by vectorization. Inf. Sci. 527, 311–328 (2020)

15. Olawoyin, A.M., Leung, C.K., Choudhury, R.: Privacy-preserving Spatio-temporal patient data publishing. In: Hartmann, S., Küng, J., Kotsis, G., Tjoa, A.M., Khalil, I. (eds.) DEXA 2020, Part II. LNCS, vol. 12392, pp. 407–416. Springer, Cham (2020). https://doi.org/10.1007/978-3-030-59051-2_28

16. Leung, C.K.-S., Jiang, F.: Big data analytics of social networks for the discovery of "following" patterns. In: Madria, S., Hara, T. (eds.) DaWaK 2015. LNCS, vol. 9263, pp. 123–135. Springer, Cham (2015). https://doi.org/10.1007/978-3-319-22729-0_10

17. Souza, J., Leung, C.K., Cuzzocrea, A.: An innovative big data predictive analytics framework over hybrid big data sources with an application for disease analytics. In: Barolli, L., Amato, F., Moscato, F., Enokido, T., Takizawa, M. (eds.) AINA 2020. AISC, vol. 1151, pp. 669–680. Springer, Cham (2020). https://doi.org/10.1007/978-3-030-44041-1_59

18. Jiang, F., Leung, C.K.-S.: Stream mining of frequent patterns from delayed batches of uncertain data. In: Bellatreche, L., Mohania, M.K. (eds.) DaWaK 2013. LNCS, vol. 8057, pp. 209–221. Springer, Heidelberg (2013). https://doi.org/10.1007/978-3-642-40131-2_18

19. Leung, C.K.-S., MacKinnon, R.K.: Balancing tree size and accuracy in fast mining of uncertain frequent patterns. In: Madria, S., Hara, T. (eds.) DaWaK 2015. LNCS, vol. 9263, pp. 57–69. Springer, Cham (2015). https://doi.org/10.1007/978-3-319-22729-0_5

20. Leung, C.K.-S., MacKinnon, R.K.: BLIMP: a compact tree structure for uncertain frequent pattern mining. In: Bellatreche, L., Mohania, M.K. (eds.) DaWaK 2014. LNCS, vol. 8646, pp. 115–123. Springer, Cham (2014). https://doi.org/10.1007/978-3-319-10160-6_11

21. Leung, C.K.-S., Tanbeer, S.K.: Mining popular patterns from transactional databases. In: Cuzzocrea, A., Dayal, U. (eds.) DaWaK 2012. LNCS, vol. 7448, pp. 291–302. Springer, Heidelberg (2012). https://doi.org/10.1007/978-3-642-32584-7_24

22. Aggarwal, C.C., Reddy, C.K.: Data Clustering: Algorithms and Applications (2014)

23. El Malki, N., Cugny, R., Teste, O., Ravat, F.: A new accurate clustering approach for detecting different densities in high dimensional data. In: Golfarelli, M., Wrembel, R., Kotsis, G., Tjoa, A.M., Khalil, I. (eds.) DaWaK 2021. LNCS, vol. 12925, pp. 167–179. Springer, Cham (2021). https://doi.org/10.1007/978-3-030-86534-4_16

24. Kim, J., et al.: KNN-SC: novel spectral clustering algorithm using k-nearest neighbors. IEEE Access 9, 152616–152627 (2021)

25. Simovici, D.A.: CLUSTERING: Theoretical and Practical Aspects (2021)

26. Braun, P., et al.: Game data mining: clustering and visualization of online game data in cyber-physical worlds. Procedia Comput. Sci. 112, 2259–2268 (2017)

27. Hoque, M.N., et al.: Reframing in clustering. In: IEEE ICTAI 2016, pp. 350–354 (2016)
28. Lee, R.C., et al.: An innovative majority voting mechanism in interactive social network clustering. In: ACM WIMS 2017, pp. 14:1–14:10 (2017)
29. Ben HajKacem, M.A., Ben N'Cir, C.-E., Essoussi, N.: Spark based text clustering method using hashing. In: Golfarelli, M., Wrembel, R., Kotsis, G., Tjoa, A.M., Khalil, I. (eds.) DaWaK 2021. LNCS, vol. 12925, pp. 137–142. Springer, Cham (2021). https://doi.org/10.1007/978-3-030-86534-4_12
30. Singh, S.P., et al.: Analytics of similar-sounding names from the web with phonetic based clustering. In: IEEE/WIC/ACM WI-IAT 2020, pp. 580–585 (2020)
31. Dierckens, K.E., et al.: A data science and engineering solution for fast k-means clustering of big data. In: IEEE TrustCom-BigDataSE-ICESS 2017, pp. 925–932 (2017)
32. Froese, R., et al.: The border k-means clustering algorithm for one dimensional data. In: IEEE BigComp 2022, pp. 35–42 (2022)
33. Nelson, J.D.: On k-means clustering using Mahalanobis distance. MS thesis, NDSU, USA (2012)
34. Fan, T., et al.: Correlation-aware sport training evaluation for players with trust based on Mahalanobis distance. IEEE Access **10**, 16898–16905 (2022)
35. Forgy, E.: Cluster analysis of multivariate data: efficiency versus interpretability of classifications. Biometrics **21**, 768–780 (1965)
36. MacQueen, J.: Some methods for classification and analysis of multivariate observations. In: Berkeley Symposium on Mathematical Statistics and Probability, pp. 281–297 (1967)
37. Likas, A., et al.: The global k-means clustering algorithm. Pattern Recogn. **36**(2), 451–461 (2003)
38. Dunford, N, Schwartz, J.T.: Linear Operators, Part 1: General Theory (1988)
39. Steinley, D.: Properties of the hubert-arable adjusted rand index. Psychol. Methods **9**(3), 386–396 (2004)
40. Rosenberg, A., Hirschberg, J.: V-Measure: a conditional entropy-based external cluster evaluation measure. In: EMNLP-CoNLL 2007, pp. 410–420 (2017)

Grapevine Phenology Prediction: A Comparison of Physical and Machine Learning Models

Francisco J. Lacueva-Pérez[1]([✉]) [iD], Sergio Ilarri[2] [iD], Juan J. Barriuso[3] [iD],
Joaquín Balduque[3] [iD], Gorka Labata[1] [iD], and Rafael del-Hoyo[1] [iD]

[1] Instituto Tecnológico de Aragón, 50018 Zaragoza, Spain
{fjlacueva,glabata,rdelhoyo}@itainnova.es
[2] Department of Computer Science and Systems Engineering, University of Zaragoza,
I3A, 50018 Zaragoza, Spain
silarri@unizar.es
[3] Department of Agricultural and Natural Environment Sciences,
University of Zaragoza, 50018 Zaragoza, Spain
{barriuso,jbalduque}@unizar.es

Abstract. The reduction of plant pest treatments contributes to a more
sustainable agriculture. However, to be effective, the application of these
treatments must be performed at the correct phenological stage of the
plants. In this paper, we present the comparison of physical and ML
models to predict the phenological stage of vineyards. The performance
of both shows an average R^2 above 0.94. However, the physical models do
not generalize well and they cannot be easily improved by the inclusion
of new datasets as ML models do.

Keywords: Grapevine · Phenology prediction · Machine learning ·
IoT

1 Introduction

Achieving a sustainable agriculture requires a change in farming practices. The
disruptive fusion of technologies, such as the Internet of Things (IoT), Big Data
and Analytics (BDA), and Artificial Intelligence (AI) applied to agriculture and
live-stocking is known as Smart Farming (SF). SF follows the Industry 4.0 concept and it can boost effective innovative actions towards more productive and
sustainable agricultural practices [13].

The treatment of pests is one of the practices that more easily can show the
benefits of application of SF for increasing farming sustainability because of its

This research is partially funded by: *GRAPEVINE* project, co-financed by the European Union's Connecting Europe Facility (CEF) Grant No. INEA/CEF/ICT/A2018/
1837816. We also thank the support of the project PID2020-113037RB-I00/AEI/
10.13039/501100011033 and the funds of the Government of Aragon to research groups:
COSMOS with reference T64_20R and T17_20R IODIDE with reference T17_20R.

R. Wrembel et al. (Eds.): DaWaK 2022, LNCS 13428, pp. 263–269, 2022.
https://doi.org/10.1007/978-3-031-12670-3_24

direct impact on ecosystems. However, to be effective, most of the treatments should be applied at a given phenological stage [4].

As a first step to create pest risk prediction models, in this paper we focus on predicting the phenological stages of grapevines and compare physical models tuned by agronomists and ML models created by data scientists. Both kinds of models are created in a real scenario and exploit different types of available data. They predict eight phenological stages at which different types of pests (mildew, powdery mildew, lobesia and botritis) can more heavily affect grape yield quality and quantity.

The structure of the rest of this paper is as follows. Firstly, in Sect. 2, we provide a theoretical background of phenology and physical and ML prediction models. Secondly, in Sect. 3, we present the experimental evaluation that we performed to evaluate and compare both types of models: we describe the geographical locations of the grapevines used in our work, the datasets used, and the physical and ML models that we created, along with the evaluation of the quality of the results they provide. Then, in Sect. 4, we reflect on the results obtained to summarize the findings and conclusions, as well as to determine opportunities of improvement.

2   Background of Phenology and Prediction Models

Phenology studies the relations between the life cycles of living beings, such as plants, and environmental changes (e.g. climatic conditions, location, etc.). Traditionally, phenology physical models rely on linear functions which only consider the relation between the phenology stages and the temperature [22]. For vines, the two preferred prediction models are the *Growing Degree Day index* (*GDD*) [3] and the *Winkler Index* (*WI*) [2].

More complex models were recently created. Generalized sigmoid models for the prediction of three phenological stages (budburst, floration and veration) were reported in [16]. Four models for predicting the remaining days until the start of four phenological stages (budbreak, bloom, veraison, and harvest maturity) were created by [18]. These models applied multiple regression on multiple factors (temperature, soil degree days, solar radiation and photoperiod).

Nowadays, researchers can take advantage of BDA technologies for the automation of the data capturing and processing of multi-sourced data [5]. Thus, BDA technologies expand the number of variables that can be used by the models, benefiting both physical and ML-based models and improving their quality [8].

In particular, ML techniques can be used to better understand the structure of time series, reduce the number of variables to be considered [19], or create synthetic data for increasing the available training datasets [15]. Besides, these techniques are used to extract patterns that can be used to classify information [14].

In the context of our work, BDA and ML are applied to determine the phenology evolution of plants [24] considering different data sources, such as field observations, field location and climatic data. These technologies also allow the automatic tuning of the hyperparameters of the models; an example is provided in [9], where manually-tuned models are compared with the ones obtained using AutoML.

3   Experimental Set up and Evaluation

Our experiments compare the predictions made by the adaptation of WI models to the areas of interest and the ML ones. The areas of interest are four wine Protected Origin Denominations (POD): Calatayud, Campo de Borja, Cariñena and Somontano. These PODs are in Aragón, a region in the north-east of Spain. The altitude of their vineyards varies from 300 meters over sea level to 1000 meters.

Four grape varieties (Mazuela/Cariñena, Grenache, Tempranillo and Chardonnay) are considered. Based on the phytopathological sensitivity of grapevines eight key stages from the BBCH scale [12] were selected: 11, 15, 63, 65, 68, 71, 77 and 81.

Several data sets are considered by our models: the Spanish Cadastral Registry [21] (providing the geographical description of the fields), the Aragón Open Data CAP registry [6] (linking the parcels to their varieties), the Red FARA repository (that provides the timely observations of the phenology stages and the pests) [7], and the SIAR database [20] (which offers the climatic station measurements). Due to the limitation imposed by the deployment of the stations of SIAR, we only considered data since the year 2009. We used data from the stations located at the PODs: the station of Borja (-1.507, 41.855) for Campo de Borja; the station at (-1.614620, 41.362900) for Calatayud; the station in Almonacid de la Sierra (-1.329960, 41.452080) for Cariñena; and the station of Barbastro (0.112610, 42.013470) for the POD of Somontano.

3.1   Raw Data Pre-processing and Analysis

We transformed the data contained in the previously introduced datasets with the guidance of agronomists and applying the DST methodology [11]. Besides, the Red FARA's phenological stage annotations were reduced to the eight stages of interest: the twenty five stages having records were aggregated in the required eight by grouping all the intermediate stages between two of them (e.g., we consider the observations of the stages 12, 13 and 14 as observations of stage 11).

Then, we took advantage of available BDA capabilities to make several transformations in all the available climatic variables: temperature, precipitation, relative humidity, solar radiation, wind speed and direction. More relevant transformations, as ML models variables selection later showed, are those related with temperature: the calculation of the force units, the daily contributions to the GDD and WI indexes, considering daily data (the conventional way) but also the 30-min measurements provided by SIAR stations; the use of different base temperatures (4.5 °C and 10 °C) in the calculation of the daily contributions; the consideration of the influence of temperatures higher than 30 °C; and the accumulation of the daily contributions to the GDD and WI considering the conventional dates (January the 1st and February the 1st), but also the start of a new crop season (the first day of autumn having a maximum temperature lower than 10 °C, when the grapevine enters into the dormancy stage); finally we considered the influence of temperatures in the development of grapevines by the calculation of the chilling cumulative hours using the models described in [17,23].

The resulting dataset is composed of 94 different attributes with values of daily data provided by the climatic stations and the different indexes obtained considering the different combinations of the parameters in the previous list. With this dataset we started to analyze its content and to create the models.

3.2   Experimental Results of the Physical Models

The objective of the physical models was to create one for each variety of interest based on the WI. These models were created using data from the field observations of the phenology and the temperature provided by the stations of the SIAR network during the period from 2009 to 2018. Furthermore, the data of 2019 was reserved for validation. For the calculation of the GDD and WI indexes of each year, a base temperature of 10 °C and the accumulation of forcing units started on February 1st were considered. The GDDs and WIs indexes for each DOP were calculated and then joined with the phenology field observations of each variety and year. The comparison of the merged data sets showed that the WI had more solid behavior for the varieties and the areas of interest.

Linear regressions were applied to the WI on each of the resulting variety datasets for adapting the WI to the varieties of interest. The four models obtained performed well achieving values of R^2 ranging from 0.92 for Tempranillo to 0.97 for Cariñena.

With the models obtained, the intervals of the WI at which each of the phenological stage of the BBCH scale will be present were determined. Moreover, given a known value of the WI, it is possible to know on which day the BBCH would be achieved. However, this would require using weather forecasts and the performance of the model would depend on the accuracy of those forecasts.

3.3   Experimental Results of the ML Models

The first ML phenology prediction models were created in three iterations on the available data for the period 2015–2019, reserving 2018 data for validation. One model per phenological stage of interest was created. These models predict the number of remaining days needed from a date to reach the phenological stage of the model. Moreover, the models were tuned to obtain a more accurate result in the period between seven and twenty one days from the prediction day. The hyperparameters were tuned using Optuna [1] during 250 trials.

We used the Random Forest (RF) algorithm for obtaining our first 8 ML models because of its explainability capabilities, although its performance is rather low (resulting models provided an R^2 of 0.8). For the second iteration, we trained the new 8 models using an ANN (Artificial Neural Network) algorithm. For optimizing the hyper-parameters of the models obtained we used Optuna during 250 trials. As expected, these models performed better than the RF ones: they had an average R^2 of 0.91.

For stages 63 and 65, the R^2 decreased until 0.78. On average, these stages respectively occur around the 149th and the 155th day of the year. They can happen in 6 days whereas two consecutive field observations are available in a period between 4 and 12 days. In consequence, both stages can overlap, making them indistinguishable for the ML models. Thus, it was decided to merge both stages and to create a new ANN-based model. This increased the R^2 for stage 63 from 0.88 to 0.97. The final values of R^2 range from 0.91 for the BBCH stage 11 to 0.97 for the stage 68, having an average R^2 of 0.96.

4   Conclusions and Future Work

This paper compared the results obtained from the tuning of the WI to the data obtained from four wine PODs and the models obtained by the application of RF and ANNs algorithms on these data. We created a physical model for each variety of interest. Eights ML models were created, one for each of the phenology stages of interest. Both kinds of models present similar values of R^2, which are over 0.94 on average.

To be able to predict, the physical methods depend on the weather forecast. This links their accuracy to the precision of the forecast (currently the weather can be predicted, with reasonable precision, no more than three days in advance). Our ML models do not consider weather forecast data and they are optimized for a time horizon between two and 3 weeks, which is the required time to be useful for a pest prediction model. Because time (the number of days to stage) is the dimension to predict, the accuracy of the models will be limited by the dataset with a lower time resolution: the field observations.

Besides, the physical models must be fine-tuned for the specific geographical areas where they are going to be applied. For example, the results for the wine variety Tempranillo show this: its R^2 ranges from 0.92 in DOP Calatayud to 0.97 in Campo de Borja. This could be solved by increasing the number of phenology field observations, which requires one year collecting new data, or by considering other new types of datasets (i.e., additional variables).

Although the R^2 metric is easy to be calculated for physical and ML models, other metrics, such as F1, recall or precision will provide better insights about model performance when considered as classifiers.

The use of the DST methodology, for guiding and documenting the model development processes, will allow us to replicate it for further improvements of ML models and also to create the pest models. Following the road-map we defined in [10], we are currently working on the improvement of the ML models by including additional data sets that contain weather forecasts and data derived from Sentinel 2 hyperspectral images. Then, we will create the pest models.

References

1. Akiba, T., Sano, S., Yanase, T., Ohta, T., Koyama, M.: Optuna: a next-generation hyperparameter optimization framework. In: 25th ACM SIGKDD International Conference on Knowledge Discovery & Data Mining, pp. 2623–2631 (2019)
2. Amerine, M., Winkler, A.: Composition and quality of musts and wines of California grapes. Hilgardia 15(6), 493–675 (1944)
3. Bonhomme, R.: Bases and limits to using 'degree. day' units. Eur. J. Agron. 13(1), 1–10 (2000)
4. Eichhorn, K.W., Lorenz, D.H.: Phenological development stages of the grape vine. Nachrichtenblatt Deutschen Pflanzenschutzdienstes 29(8), 119–120 (1977)
5. Fenu, G., Malloci, F.M.: Forecasting plant and crop disease: an explorative study on current algorithms. Big Data Cogn. Comput. 5(1), 2 (2021)
6. Government of Aragón: Aragón Open Data Home Page (2022). https://opendata.aragon.es/. Accessed 30 May 2022
7. Government of Aragón: Red FARA Home Page (2022). http://web.redfara.es/. Accessed 30 May 2022
8. Kamilaris, A., Kartakoullis, A., Prenafeta-Boldú, F.X.: A review on the practice of big data analysis in agriculture. Comput. Electron. Agric. 143, 23–37 (2017)
9. Kasimati, A., Espejo-García, B., Darra, N., Fountas, S.: Predicting grape sugar content under quality attributes using normalized difference vegetation index data and automated machine learning. Sens. (Basel Switz.) 22(9), 3249 (2022). https://doi.org/10.3390/s22093249
10. Lacueva-Pérez, F.J., Artigas, S., Vargas, J., Lezaun, G., Alonso, R.: Multifactorial evolutionary prediction of phenology and pests: can machine learning help? (2020). https://doi.org/10.5220/0010132900750082
11. Martinez-Plumed, F., et al.: CRISP-DM twenty years later: from data mining processes to data science trajectories. IEEE Trans. Knowl. Data Eng. 1 (2019). https://doi.org/10.1109/TKDE.2019.2962680
12. Meier, U., et al.: The BBCH system to coding the phenological growth stages of plants-history and publications. J. Kult. 61(2), 41–52 (2009)
13. Moysiadis, V., Sarigiannidis, P., Vitsas, V., Khelifi, A.: Smart farming in Europe. Comput. Sci. Rev. 39, 100345 (2021). https://doi.org/10.1016/j.cosrev.2020.100345
14. Najafabadi, M.M., Villanustre, F., Khoshgoftaar, T.M., Seliya, N., Wald, R., Muharemagic, E.: Deep learning applications and challenges in big data analytics. J. Big Data 2(1), 1–21 (2015). https://doi.org/10.1186/s40537-014-0007-7
15. Rasheed, A., San, O., Kvamsdal, T.: Digital twin: values, challenges and enablers from a modeling perspective. IEEE Access 8, 21980–22012 (2020)
16. Reis, S., et al.: Grapevine phenology in four Portuguese wine regions: modeling and predictions. Appl. Sci. 10(11), 3708 (2020)
17. Richardson, E.A.: A model for estimating the completion of rest for 'Redhaven' and 'Elberta' peach trees. HortScience 9, 331–332 (1974)
18. Schrader, J.A., Domoto, P.A., Nonnecke, G.R., Cochran, D.R.: Multifactor models for improved prediction of phenological timing in cold-climate wine grapes. HortScience 55(12), 1912–1925 (2020)

19. Sirsat, M.S., Mendes-Moreira, J., Ferreira, C., Cunha, M.: Machine learning predictive model of grapevine yield based on agroclimatic patterns. Eng. Agric. Environ. Food **12**(4), 443–450 (2019). https://doi.org/10.1016/j.eaef.2019.07.003
20. Spanish Ministry of Agriculture, Fisheries and Food: SIAR Home Page (2022). https://eportal.mapa.gob.es//websiar/Inicio.aspx. Accessed 30 May 2022
21. Spanish Treasury: Spanish Cadastral Registry Electronic Home Page (2022). http://www.catastro.minhap.es/webinspire/index.html. Accessed 30 May 2022
22. Wang, Y., Case, B., Rossi, S., Dawadi, B., Liang, E., Ellison, A.M.: Frost controls spring phenology of juvenile Smith fir along elevational gradients on the southeastern Tibetan Plateau. Int. J. Biometeorol. **63**(7), 963–972 (2019). https://doi.org/10.1007/s00484-019-01710-4
23. Weinberger, J.H.: Chilling requirements of peach varieties. J. Am. Soc. Horticult. Sci. **56**, 122–128 (1950)
24. Wolfert, S., Ge, L., Verdouw, C., Bogaardt, M.J.: Big data in smart farming – a review. Agric. Syst. **153**, 69–80 (2017). https://doi.org/10.1016/j.agsy.2017.01.023

Author Index

Printed in the United States
by Baker & Taylor Publisher Services